ABOLISH OIL NOW!

ALSO BY ERIK D. CURREN

Buddha's Not Smiling: Uncovering Corruption in Tibetan Buddhism Today

The Solar Sales Leap: Stop Knocking on Doors, Cold Calling & Buying Leads and Start using the INTERNET to Grow Your Solar Business for the Long Term

The Solar Patriot: A Citizen's Guide to Helping America Win Clean Energy Independence

ABOLISH OIL NOW!

ABOLITIONISTS BEAT SLAVERY AND ABOLITIONISTS CAN BEAT CLIMATE CHANGE

ERIK D. CURREN

NEW SKY BOOKS

Published in the United States by New Sky Books.

Library of Congress Cataloging-in-Publication Data is available.

ISBN 978-1-7379819-0-9

New Sky Books
www.newskybooks.com

1 3 5 7 9 10 8 6 4 2

First Edition

For Lindsay, Anwyn, and Chloë

CONTENTS

Prologue on the Use of History

To be ignorant of what occurred before you were born is to remain always a child.

—Marcus Tullius Cicero

Statues have been coming down across the United States and all over the world. Good riddance to most of them. Slave traders and Confederate generals never should have been put up in the first place. Let's hope that removing statues and changing street names is not the end, but the beginning, of a new push to make America and all other countries fair for all citizens regardless of race.

In our quest for social justice, we should remember that history is populated not only by villains but also by heroes. Black and white activists who fought to end slavery, which had plagued human societies for thousands of years, were called abolitionists. Their idealism and persistence succeeded against what everybody at the time agreed were impossible odds.

Today, the movement to save the nations of the earth from climate disaster also faces long odds. But not as long odds as abolitionists faced to end slavery. That's the good news. It's encouraging that solutions are here. And it's frustrating that we've had those solutions for years. I know that because I work in solar power. We could have started on the path to climate safety 30 or 40 years ago. But we didn't. That's because big fossil fuel companies blocked us.

People who care about the climate could have learned from history how to unblock progress. Without history, we've been fighting with one arm tied behind our back. We need help from abolitionists back then so that we can become abolitionists too. That's how we'll win.

Then, maybe we can start putting up some better statues for future generations to admire.

Introduction

I am aware that many object to the severity of my language; but is there not cause for severity? I will be as harsh as truth, and as uncompromising as justice. On this subject, I do not wish to think, or to speak, or write, with moderation. No! no! Tell a man whose house is on fire to give a moderate alarm; tell him to moderately rescue his wife from the hands of the ravisher; tell the mother to gradually extricate her babe from the fire into which it has fallen;—but urge me not to use moderation in a cause like the present. I am in earnest—I will not equivocate—I will not excuse—I will not retreat a single inch—AND I WILL BE HEARD.

—William Lloyd Garrison, *The Liberator*, 1831

I think history will look back on the covert operation run by the fossil fuel industry against its own country to debilitate and incapacitate our own government from addressing this problem as one of history's vilest political acts.

—Senator Sheldon Whitehouse (D-RI), Congressional hearing, 2021[1]

This book compares burning fossil fuels to human slavery. Both helped build the American economy, for good and for ill, making our country rich and powerful. Slavery was a low-cost source of work, but it was oppressive and cruel to Africans and Black Americans.[2] Slavery also indirectly harmed many white Americans. Fossil fuels are a low-cost source of energy but they're dangerously polluting, which has also made them oppressive and cruel to everybody, but in a tragic irony, especially to Black and brown people.

The negatives of slavery always outweighed its supposed positives. At first, abolitionists in the North and tobacco and cotton planters in the South alike knew that slavery was wrong and hoped that it would fade away in the future. But once the profits of producing crops with bound labor started to rise more quickly, the two groups parted ways. Even while abolitionists continued to fulminate against holding people in chains for no reason other than the color of their skin, later generations of Southern planters seemed to forget how bad slavery was. As the big money began to get even bigger, those planters even started to justify slavery as a good arrangement for everybody, including the Black man or woman in chains.

It's the same with fossil fuels. It didn't take climate change for people to realize that it was wrong to build the economy on an energy source that was always deadly and dangerous. In 1285, after rich people burned so much coal for heat that the air in London became a brown haze, King Edward I of England set up the world's first commission to cut air pollution and banned the burning of coal. The authorities in western countries have known for more than 700 years that fossil fuels were dangerous. But when the money started to get good from mining coal, then coal barons started to act like the pollution was no big deal. King Edward's coal ban didn't last long.

Today we know that no country that wanted to be happy, prosperous, and free should ever have depended on either slavery or dirty energy. There's no changing the past, but we must be aware of the past to do better today. Now we must take control of our future. And if it took the risk of climate collapse to help us see that the benefits of fossil fuels are not now and have never been worth the costs, then let's save the climate to save our future.

The great mission of the generation before the Civil War was to abolish slavery, which they succeeded in doing in the 1860s. The great mission of our generation is to abolish fossil fuels by 2050 or even earlier.

With justified pride, our country has long celebrated the abolition of slavery. Today Americans wonder how a nation dedicated to the proposition that all people are created equal could ever have tolerated the abomination of building wealth off the backs of men and women dehumanized and treated like chattel.

We also wonder how a nation dedicated to the pursuit of happiness could still tolerate the abomination of building wealth by laying waste to towns located near fracking wells, coal ash pits, and oil refineries. We also wonder how our leaders can continue to sell out the future of our country and our world to a small cabal of dirty energy owners and investors who put their profits before their country and their fellow humans.

A book that compares oil to slavery is obviously not a moderate approach for gradual solutions. It's an immoderate approach to douse a national and global house on fire. With the grave threat of climate disruption staring down world civilization, I insist that nothing less will do.

No Precedent Other than Abolition

"Climate change and ecological breakdown may one day be viewed with the same universal repugnance as slavery," British broadcaster and naturalist David Attenborough told a committee of members of Parliament in 2019. "There was a time in the 19th century when it was perfectly acceptable for civilized human beings to think it was morally acceptable to actually own another human being as a slave. And somehow or other in the space of I suppose 20 or 30 years the public perception of that totally transformed. By the middle of the 19th century, it was becoming intolerable." When it comes to the environment, Attenborough said, "we are right now in the beginning of a big change."[3]

Of course, for certain people it was never acceptable to own a slave. Those people were called abolitionists.

The first abolitionists were enslaved Africans and their descendants. Later, white people of conscience came along to help. Even when you add up both groups, abolitionists were never a majority of the population—not even close. But through persistent and dedicated action, this vocal minority succeeded in moving the majority to act to end slavery.

The good news today is that climate solutions are much more popular than abolition ever was. Americans overwhelmingly support cutting the pollution that's blanketing the Earth and switching from fossil fuels to clean energy.[4] That gives us a huge advantage today over abolitionists back then, who were some of the most unpopular people in the country, both North and South.

Also, slavery was about people, and so it became about emotional issues in politics and culture like race, prejudice, tradition, and power struggles among different groups. When it comes to the biggest climate solution, switching the economy from dirty energy to clean energy, it's just about technology and economics, rather than how a person's appearance or parentage affects their position in society.

Finally, renewable energy is so superior to fossil fuels in nearly every practical way, if you are going to get emotional about it, it's much easier to love a clean solar panel or wind turbine than a dirty oil well or coal mine.

It costs money to build any power plant, whether to burn natural gas or soak up the sun. But once the equipment is in place, the difference in cost couldn't be more obvious. Power plants that burn fossil fuels must keep buying supplies of fossil fuels. But solar panels and wind turbines once built don't have to pay anything for sunlight or moving air. The fuel for renewable power is free.

Does that even need to be said twice?

When you think about it this way, it's hard to understand why we still burn any fossil fuels at all instead of just spending the money to build enough wind turbines and install enough solar panels to start using free fuel. Even if you don't care about pollution at all the money should speak for itself.

There's a simple reason why clean energy isn't all over the place, and it's not because of technology or cost. The reason we don't have more solar and wind and we still use so much oil, gas, and coal is purely political. People whose wealth is invested in dirty energy stand to lose a lot of that wealth if the U.S. economy adopts clean energy. Those wealthy people use their money to buy political power to block clean energy. So far, they've succeeded all too well at this shameful game, which has helped keep them rich but has put everyone's future in peril.

That's why we need abolition. To slow and stop climate change, we must defeat the most powerful moneyed special interest of our day. That's exactly what abolitionists faced in their campaign to abolish slavery, the most powerful moneyed interest of their day.

For abolitionists, it was the sugar and cotton growers and their allies among merchants, factory owners, bankers, and shippers.

For people who want to save the climate the adversaries are oil, gas, and coal producers along with Wall Street banks, plastic factories, pipeline construction companies, and of course attorneys, lobbyists, and PR guys. Lots of those. Together, this Axis of Oil has dominated American politics for half a century or more. It's true that fossil fuel companies have been weakened in recent years by economic and political setbacks. But the dirty energy lobby remains among the most powerful combinations of wealth and power on earth. It will require a formidable movement of people power to go up against their money power.

Sounding the same note as naturalist David Attenborough, leading writers and activists on the climate from Al Gore to Bill McKibben to

Naomi Klein have invoked the movement to end slavery in the eighteenth and nineteenth centuries as an apt analogy for the huge, transformative movement needed to save the world from catastrophic global heating today.

"The climate movement should be seen in the context of the great moral causes that have transformed and improved the outlook for humanity," Al Gore has said. "It was wrong to allow slavery to continue, it was wrong to deny women the right to vote, it was wrong to discriminate on the basis of skin color or who you fell in love with."[5]

Naomi Klein agrees that climate is as strong a moral issue as slavery was. She also thinks that the success of abolitionists in fighting the big money behind slavery provides a powerful model to citizens who must go up against the big money of fossil fuels to save the climate. "Abolition succeeded in challenging entrenched wealth in ways that are comparable to what today's movements must provoke if we are to avert climate catastrophe," writes Klein.[6]

Only abolition provides a historical precedent for the huge scale of wealth that activists are hoping to force a moneyed special interest to renounce, in the form of fossil fuels left in the ground, according to MSNBC broadcaster Chris Hayes. "The climate justice movement is demanding that an existing set of political and economic interests be forced to say goodbye to trillions of dollars of wealth. It is impossible to point to any precedent other than abolition."[7]

The scale of change required to avoid climate catastrophe will be huge, as it was to abolish slavery. It's not just about technical fixes in international treaties or environmental regulations to cut a few emissions here and there.

"If we are serious about climate change, we need to dismantle the fossil fuel economy and replace it with a moral economy that values ecosystems, sufficiency, distributive justice, and real democracy. And that kind of transformation will not come without struggle. The only precedent that comes close in scope is the movement to dismantle the slave economy: the abolitionist movement," according to Denise Fairchild of the Urban Resilience Project. Writing that all Americans are slaves to the fossil fuel economy today, Fairchild argues that only a movement as big as abolition will transform four areas that we need to free ourselves: property rights, profits, privilege, and power.

Abolitionists successfully challenged the idea that some people were property who could be bought and sold. Climate activists need to dislodge the idea that nature is property to be used primarily for profit while also challenging the free-market orthodoxy that property rights trump the human right to a livable planet.

As to profits, before the Civil War, about 3,000 wealthy planters owned most slaves in the American South. The other eight million white people who lived in slave states at the time got little or no direct benefit from slavery but were harmed by slavery both economically and politically. Big slave planters monopolized agriculture, crowding out small farmers. Those same elite planters dominated politics in the South, enforcing gag rules on abolitionists that chilled free speech on other topics across the political spectrum. So, ending slavery would have helped most Southerners, Black and white.

Yet, after slavery was over, Northern industrialists swooped in to make massive profits during the Gilded Age, leading to a new set of inequities between capital and labor. Today, as we transition from a dirty economy to a clean one, which will involve a huge transfer of wealth, we must ensure that a new oligarchy does not arise and claim most of the economic benefits for itself. The prosperity that the clean economy will create must be distributed more fairly than wealth has been in the past.

Finally, power must not be used to bolster unfair privilege. The U.S. Constitution, for example, denied Black people citizenship until the passage of the Fourteenth Amendment in 1868. Though women of all races would have to wait until 1920 for the franchise, Black men got the right to vote with the Fifteenth Amendment in 1870. Even after that, many decades of activism were still required to protect civil rights, a fight that continues today. The climate justice movement must target the power centers in finance and government that allow fossil fuel companies to exploit people and damage the planet.[8]

A monumental effort will be needed to make the necessary changes but making an analogy of fossil fuels to slavery may sound overdramatic or insensitive to some readers.

Starting with the dawn of civilization, slavery was practiced by peoples from the ancient Israelites, Greeks, and Romans to the tribal kingdoms of early medieval Europe to North African pirates in the nineteenth century. Enslaved people came from different races and ethnic backgrounds but often masters and slaves didn't look very different from each other.

But in the Western Hemisphere since Columbus, and in North America since the beginning of British settlement at Jamestown, slavery was always about race. After the initial attempt by Europeans to enslave indigenous people failed early on, slavery's suffering was primarily inflicted on people of African descent. This racial difference between the enslaver and the enslaved made American slavery harsher to endure and harder to escape than previous forms of slavery found elsewhere. Whether in ancient Greece or medieval Europe, since slave and master were usually of the same race, slaves were treated with more respect, and it was easier for them to gain their freedom. Then, once freed, it was easier to blend into the larger society.

By contrast, in the Western Hemisphere, the paths to freedom in white society for enslaved Black people were fewer and those paths were blocked by more obstacles. Once freed from slavery, Black Americans faced obstacles put up by white supremacy at every turn.

A century and a half after slavery's legal demise in the United States, the oppressive system of treating people like property based on their race continues to extend its cold, dead hand from beyond the grave to cast a shadow of prejudice, discrimination, poverty, and violence over Black Americans today. Through centuries of suffering, those Americans have earned a special moral stake in this history, a right like the one that Jewish people everywhere have earned to talk about the Holocaust.

We should be careful about comparing any modern-day catastrophe, no matter how bloody and brutal, to either the Holocaust or to slavery. "It's almost always foolish to compare a modern political issue to slavery, because there's nothing in American history that is slavery's proper analogue," writes Chris Hayes. In saying that the climate movement needs to become the new abolitionism, he is not making a moral equation between the enslavement of Black people and the burning of hydrocarbons to run machines. "Humans are humans; molecules are molecules."

But it is valid to compare the political economy of slavery with that of fossil fuels. The money in each case is so huge as to present an almost insurmountable barrier to putting the public interest ahead of the interests of a few rich people who will fight to the death to keep their money.

We've learned that we should be equally careful before we compare any contemporary political leader, no matter how dangerous a demagogue he or she may be, to Hitler. We should be just as careful when we compare any business leader today to a slaveowner before the Civil War. One of the few things left in our cynical age that Americans recognize as sacred in our nation's past is the unparalleled suffering of enslaved Black people before the Civil War.

We should tread the abolition trail with care. But we should not avoid this path. Indeed, it may be the very road we need to unblock serious action on climate. And as climate justice leaders such as researcher Robert Bullard and Rev. William J. Barber have explained, racial discrimination and other aftereffects of slavery have made fossil fuel pollution worse for Black Americans and people of color in general. We'll examine this legacy as we go along.

After spending years in my own career as a volunteer climate activist promoting market-friendly, bipartisan solutions to climate change designed to be acceptable to fossil fuel companies and their political allies, I am keenly aware of another objection to this book's approach: Saying that fossil fuels are like slavery is not likely to win friends and build influence at traditional energy companies.

So let me be clear. I do not think that people who work for oil, coal, and gas companies today are evil.

Quite the opposite: I recognize their role in providing reliable energy to our country and the world for more than a century. The roster of intelligent, creative, and conscientious people in fossil fuel companies is deep. Even more, some people inside of oil, gas, and coal companies recognize the danger of climate disruption and want to help protect human civilization from that danger. Many employees of these companies, especially younger ones, wish that their companies would do less damage. There are even people inside the industry who want it to transition to solar and wind power.

Fossil fuel workers are clearly not as personally involved in oppression as were slave-owning cotton planters, slave-driving overseers, and captains of slave ships plying the hellish Middle Passage from Africa across the Atlantic to the Americas in past centuries. Lately, fossil fuel companies have started putting out messages that they have changed their tune on climate. Oil companies now say that they want to be part of the solution rather than the cause of the problem. They're promising to achieve "net-zero" climate emissions in the next few decades.

A more conciliatory approach would certainly go down better in a boardroom at the ExxonMobil headquarters in Houston, in the office of a chemical plant up the Mississippi River from New Orleans, or on the floor of the United States House of Representatives. And some climate experts think that those are exactly the places we need to reach to make any difference.

The Contributions of Traditional Energy

"I think we need to forgive fossil fuels," says Saul Griffith, who won a "Genius" grant from the MacArthur Foundation in 2007 for his work as an inventor in optics, textiles, nanotechnology, and energy analysis. "We are demonizing them right now. We are getting the fossil fuel industry sort of backed into a corner. And honestly, the people in this country who know [best] how to build infrastructure at scale are in the fossil fuel industry."[9]

Griffith is half right. Though dirty energy's role in a rapidly heating world must continue to shrink, it's only fair to recognize fossil fuels for their past contributions to building the industrial and information economies that have made the United States and other advanced economies so prosperous and delivered innovations to improve the lives of billions of people.

Coal miners, construction workers who build pipelines, engineers, chemists, and roughnecks on offshore oil platforms should not be blamed for the climate crisis or for the air and water pollution that their companies' facilities have inflicted on local communities. Working men and women in oil, gas, and coal have applied their brawn and brainpower alike to provide energy for the rest of us for two centuries. These workers have made sacrifices: Coal miners have suffered black lung disease and oil workers have died in explosions of wells.

We should honor the people of the traditional energy industry and recognize the contributions that fossil fuels have made to human society and especially to the United States as the world's largest producer and user of energy. Concentrated and reliable energy from fossil fuels was the only force available at the time at scale to power the Industrial Revolution. Ever since, abundant energy has provided untold benefits. For all the problems that we recognize today, we should remember that coal, oil, and gas have each played a crucial role in creating the modern world. It's not the workers' fault that pulling carbon out of the ground has always been so dirty and so dangerous. If anything, workers themselves on the front lines of mining and drilling suffered the most.

As dirty as fossil fuels always were, they also offered their own benefits to the environment. It seems ironic today, but when it hit the scene in the eighteenth century, coal helped save forests from being cut down for fuel and provided a more efficient way for millions of families to heat their homes in the winter.

And talk about building wealth. Coal ran the railroads that tied modern nations together and the steamships that boosted world trade. Coal powered factories that turned iron into bridges that spanned rivers and girders that held up skyscrapers. Coal ran power plants that made electricity affordable enough to light up the cities of the world. Coal enlisted as America's ally in war and stepped up as our friend in peace, providing well paid union jobs that helped lift millions of families into the middle class.

Half a century later, oil stepped up to run even faster trains and even bigger ships. Later, oil powered the transportation revolution that gave us roads, cars, buses, and trucks, cutting travel times for people and goods from months and weeks to mere days and hours. Petroleum gave us chemicals to produce lifesaving drugs along with fertilizers and pesticides that enabled America's farmers to produce an unprecedented surplus of food. Oil also gave us plastics to make food packaging, clothing, and medical necessities from disposable personal protective wear for doctors and nurses to heart valves and prosthetic limbs.

Finally, in recent decades, natural gas brought electric power that was cheaper and more flexible than coal, though because of methane leaks from wells and pipelines, gas was never cleaner than coal as the industry has falsely claimed. Affordable electricity helped make the internet revolution possible, connecting people as never before, opening good jobs and exciting careers and building unprecedented wealth.

Civil rights advocates asking that Black Americans today receive reparations payments for the work of their enslaved ancestors estimate that before abolition, slave labor contributed trillions of dollars' worth of value to the U.S. economy.[10] After abolition, when fossil fuels replaced slaves as America's source of cheap energy, oil, gas, and coal helped build infrastructure while providing products and services also worth trillions to customers, workers, and investors.

It's right to remember and even honor the many ways that fossil fuels enabled the modern world with all its comforts and innovations, saving many lives and enhancing many more. At the same time, we cannot forget how, from the very beginning, pollution from fossil fuels has fouled our water and air and how plastics have crowded landfills and invaded the oceans. We must also debit the industry for the many clever and devious ways they have used money, lobbyists, and lies to pollute and pervert our democracy.

Though we should offer respect and gratitude to the people who made traditional energy possible, it is not time to forgive the fossil fuel industry. Instead, it is time to thank oil, coal, and gas companies for their past service—and then send them home to retirement. Or we can invite them to help with our energy transition. Many of these companies have started claiming that they want to be "part of the solution" on climate. In a promising development, big oil companies including Shell with experience in offshore oil have already started to buy leases to install offshore wind.[11]

But let's be clear on the difference between what's helpful and what's just greenwashing. It's not acceptable for fossil fuel companies to find clever new ways to use oil, gas, and coal, as these companies are trying to do now. They claim that such schemes as carbon capture and sequestration or using natural gas to produce hydrogen will allow them to use their remaining reserves of fossil fuels without harming the climate. This is a dubious proposition at best and a straight-up con at worst.

Being part of the solution must mean abandoning oil, gas, and coal entirely. It must mean leaving fossil fuels in the ground. And if traditional energy companies are serious about that, not only will they change their business and stop exploring for and producing oil, gas, and coal. These companies will also stop spreading propaganda and paying lobbyists to protect fossil fuel production and start spreading the truth. They'll take the bold step to begin lobbying against their current business and start lobbying for government action to force companies to leave fossil fuels in the ground.

Many people inside dirty energy companies already know that this is the only way their companies can justify continuing to do business in the future. Climate-friendly people inside traditional energy companies need help from the outside to convince CEOs, corporate boards, and investors to transform.

Whatever their companies do, justice and economics alike require that we welcome willing fossil fuel workers into the new clean energy economy. Their skills and dedication will be essential to installing enough solar panels and wind turbines and connecting those to enough batteries to replace the dirty energy that our economy has come to rely on. That's just good business. And as a matter of equity, society must not leave oil, gas, and coal workers, their families, and their communities behind.

Converts as Valuable Allies for Abolition

The climate movement should also welcome leaders of fossil fuel companies who want to help transition to clean energy, just as the abolition movement welcomed reformed slaveowners.

Take the Grimké sisters, two of fourteen children raised by a wealthy judge and slaveowner in South Carolina, for example. After converting to Quakerism and moving from Charleston to Philadelphia, Sarah and Angelina Grimké became outspoken critics of slavery. As stars on the abolitionist lecture circuit, the Grimkés thrilled audiences in New England while earning the hatred of their former neighbors in the Palmetto State and members of the white planter class across the South. During the Civil War, older sister Sarah wrote articles and gave lectures in support of Abraham Lincoln.

Two decades earlier, in 1836, her younger sister Angelina published a 36-page pamphlet entitled *Appeal to the Christian Women of the South*, urging white Southern women to embrace the antislavery cause. "I know you do not make the laws," Grimké wrote, "but I also know that you are the wives and mothers, the sisters and daughters of those who do; and if you really suppose you can do nothing to overthrow slavery, you are greatly mistaken."[12] Unfortunately, few white women in the South responded to Grimké's appeal. But, since she was a former slaveholder, her words carried special credibility for both men and women who were open to abolition in the North.

In the same way, former executives from oil, gas, coal, and plastics companies who now support a transition to clean energy can be powerful allies in the fight for climate solutions that are economically viable.

For example, after a career as a lobbyist for oil and plastics in Washington, DC, Lewis Freeman retired to rural Virginia, where he soon became a leader of a coalition of activists who successfully stopped an $8 billion natural gas pipeline in the summer of 2020. Sponsored by giant monopoly utilities Dominion and Duke, the Atlantic Coast Pipeline would have transported fracked gas across some of the most sensitive landscapes on the East Coast. This rural region hosts headwaters for three of America's most iconic rivers, the Ohio, the

Potomac, and the James, which provide drinking water to tens of millions of Americans.

To stop the pipeline, Freeman helped coordinate dozens of land conservation, environmental, and civil rights groups across three states. During a six-year campaign to stop the pipeline, Freeman used his organizing skills and insider knowledge of energy policy to communicate effectively with government regulators, coordinate with allies, and educate the public on the dangers of the pipeline project. Persistent citizen organizing finally paid off. In July 2020, after delays caused by opposition from activists helped to raise the projected cost of completion from $4.5 billion to $8 billion, corporate sponsors of the Atlantic Coast Pipeline cancelled the project.

Freeman is pleased with his group's victory. But he knows that dirty energy companies don't give up easily and that they may target his part of the world for another project in the future. That's why he's helped to maintain the anti-pipeline coalition even after it succeeded in stopping the Atlantic Coast Pipeline. "I think of ourselves as a NATO for the central Appalachian region," Freeman told me. Freeman is now helping to develop Conservation Hub, an online system which uses geographic information systems technology to monitor fossil fuel projects and help activists learn of leaks or other problems so those activists can alert environmental regulators.

And, to help his old industry to reform, Freeman appeared in the 2020 Frontline program "Plastics Wars," where he exposed greenwashing campaigns by plastic manufacturers around recycling. When it comes to protecting the environment, Freeman has earned credibility by making his actions match his words.[13]

Just as nineteenth century abolitionists needed people who knew about slavery from the inside and were willing to expose its secrets as the Grimké sisters did, so today's climate movement needs more people who know fossil fuels from the inside and are willing to buck their former industries as Lewis Freeman did. Anyone who wants serious action against dirty energy should extend a hand of friendship to genuinely pro-climate workers and executives alike from fossil fuel companies.

But it's past time for the fossil fuel business to end. Whether all at once through industry-wide collapse or gradually through a planned phase-out, the American people and the people of the world cannot afford another five or six decades of companies producing and selling oil, gas, and coal as dirty energy CEOs have planned.

Now is not the time to forgive fossil fuels—now is the time to make them stop polluting and then pay to clean up the mess. As we help willing workers to make a fair transition to clean energy, we must stop further damage from producing and burning dirty fuels.

Then, we must insist that the big shots among oil and coal company CEOs, investors, and lobbyists be held accountable for the costs to clean up the pollution they've spewed over the last fifty years or more. Dirty energy companies must pay to clean up the many messes from refineries, waste dumps, and chemical plants in communities from "Cancer Alley" in Louisiana to the coalfields of Appalachia to Richmond, California. These companies and their investors must also pay to mitigate the damage caused by the greenhouse pollution their products have released.

At the same time, even if they don't want to get into clean energy, fossil fuel workers deserve equity too. Because oil, gas, and coal companies were producing energy as part of an agreement with the rest of society, the losses of ending their business should not fall entirely on their own people. It's only fair that government use tax and regulatory policy to help fossil fuel operations to shut down with as little pain as possible to workers and their families and communities.

The trick will be to provide a smooth transition while ensuring that dirty energy production does in fact shut down. We cannot allow dirty energy companies, in the service of equity, to resort to tricks like carbon offsetting or doubtful technologies like carbon capture and storage to justify continuing to pump oil and gas and mine coal.

And it remains to be seen whether the wealthy investors and CEOs of fossil fuel companies deserve any compensation at all when society decides to phase out their business, devalue their investments in company stock, and shrink their future earnings. Of course, it will be easier politically to enact serious climate solutions if we can neutralize opposition from dirty energy lobbyists, and one way to do that is to pay oil company CEOs to walk away from their business, no matter how unfair that sounds. That's an issue we'll take up later.

Demanding Climate Justice

Nothing short of a massive program to cut fossil fuels while dealing with the cultural inequities that have allowed dirty energy facilities to pollute certain places more easily than others will be big enough to prevent catastrophe. Different plans are circulating that would be a good start, and the best plans offer a comprehensive program like the Just Recovery plan from the activist group 350.org with its five wide-ranging points:

1. Protect all workers and provide health care for all.

2. Provide relief directly to people, regardless of immigration status, and not corporate executives.

3. Prioritize community-led recovery in Black and indigenous communities, and communities of color around the country.

4. Stop fossil fuel extraction.

5. And invest in a Green New Deal that centers on racial justice.

But even if oil is as dangerous to the world's atmospheric climate today as slavery was to the world's moral climate in the past, isn't righting the wrongs of racial injustice a separate fight from saving the environment?

After George Floyd was killed by a Minneapolis police officer in May 2020, people of conscience nationwide demanded justice for Floyd and the many other Black and brown people who suffered death, injury, and intimidation from police violence in recent years. Black Lives Matter activists called for reforms in law enforcement and criminal justice or even full or partial defunding of police departments. Their call was partially answered by the guilty verdict for murder against former police officer Derek Chauvin in April 2021. But so much more work remains to be done.

Racial justice advocates have also called for environmental justice. Both civil rights leaders and climate activists have recently rediscovered what their predecessors in the 1960s and 1970s knew well: A healthy relationship between humans and nature is not just an interest of white people in the suburbs who take their Sierra Club tote bags to Whole Foods. In fact, cleaning up America's energy system may offer the biggest immediate benefit to people who live near dirty energy facilities. These neighbors are overwhelmingly people of color. In the longer term, preventing climate chaos is also a matter of equity.

Racial justice in cities and rural areas everywhere requires a stable physical environment with clean air and water. Black and brown Americans and low-income families have already paid the highest price so far from pollution and exposure to dangerous weather. It's no surprise that Black and brown Americans fear the effects of runaway climate heating and support serious climate solutions more than white people do on average, as shown by recent surveys.[14]

The oil industry has also been one of the worst offenders against racial fairness in its own employment practices. After the Civil War, while other industries welcomed or at least accepted Black laborers, oil and gas workers, especially in higher paid jobs, remained overwhelmingly white.

By 1910, in the timber industry, Black workers accounted for nearly 60 percent of the workforce in Louisiana, and 38 percent in Texas. But as late as 1940, Black workers held only 0.05 percent of jobs in oil production, and only 3 percent of positions in oil refineries. "In researching the history of oil in modern America since the 1860s to the present, oil was I think unquestionably the most racially homogeneous industry in America," Darren Dochuk, author of *Anointed with Oil: How Christianity and Crude Made Modern America*, explains. "There are clear racist patterns of organization within the industry from the very beginning."[15]

Bill McKibben has written that "racism, police violence, and the climate are not separate issues." Finding that the environmental movement has now merged with the environmental justice movement, McKibben thinks it's a good thing. "Over the years, the environmental movement has morphed into the environmental-justice movement, and it's been a singularly interesting and useful change. Much of the most dynamic leadership of this fight now comes from Latinx and African-American communities, and from indigenous groups; more to the point, the shift has broadened our understanding of what 'environmentalism' is all about." For McKibben, saving the climate is as much about social, economic, and racial equity as it is about the natural world.[16]

That makes the fight to abolish oil today into an extension of the project to end slavery and then to ensure civil rights for Black Americans.

An approach along the lines of 350.org's Just Recovery or the Green New Deal doesn't try to deal with climate separately from social and economic concerns, as too many white environmentalists tried to do in the past. Instead, new approaches integrate climate solutions with ideas to root out racial discrimination and other failures in our politics and economy that have made climate pollution possible in the first place.

A focus on social justice and racial equity is what the climate movement has been lacking, says Rev. William J. Barber, founder of the Poor People's Campaign. "At some point we got to be clear you cannot address climate issues unless you address systemic racism and voter suppression. Because you can't move the political people who are blocking the kind of legislation that you that we need, you can't. And we have to understand the same people that are against living wages that are against gay communities that are against public education that are against voting rights that are the same people that are against climate justice."[17]

As it was back before the Civil War when abolitionist William Lloyd Garrison refused to shut up about the evil of slavery, today America's national house is on fire, and we shouldn't shut up about it either. We cannot extinguish the heat in our atmosphere without removing the fuel provided by the racial discrimination and economic inequality that hobble our democracy. Abolition, which was a movement of movements that combined antislavery with women's rights, labor rights, and other progressive causes, provides a powerful model for a big-tent climate movement today. Only a movement this big will have any hope to put out the fire.

Time is running short. America and the world are burning in ways that generations alive today have never experienced before. Our nations are also drowning in floods, sweating in heatwaves, thirsting in droughts, starving in famines, and wasting from disease under the curse of a heating climate. And worse is yet to come unless we act decisively and succeed massively just as Garrison, Frederick Douglass, and so many other people of action and heart succeeded in ending hundreds of years of American slavery.

1

Why We Need
Abolition Once Again

If there is no struggle, there is no progress. Those who profess to favor
freedom and yet deprecate agitation are men who want crops without
plowing the ground. They want rain without thunder and lightning.
They want the ocean without the roar of its mighty waters. The
struggle may be a moral one or it may be a physical one, or it may
both moral and physical, but it must be a struggle. Power concedes
nothing without a demand. It never has and it never will.

—Frederick Douglass

Burning fossil fuels is the cause of climate change—and we can stop
burning fossil fuels. Step one is to stop burning fossil fuels. Step two is
to sequester the carbon you've already burned...It's not about
individual change. The big lever is policy. That's where the money
is...The lever for individuals isn't enough. The goal is not self-
purification. The goal is institutional and political change.

—Leah Stokes, political scientist and climate activist[1]

It sounds clichéd but looking at history makes it easy to see that to
win the future, we need to learn from the past. Climate scientists
call on natural history that's ancient, telling us that today's climate
risks getting hotter than any humans have ever faced. There's more
carbon dioxide in the atmosphere today than at any time in the last
three million years.

In an alternate universe, once the presidents and prime ministers of
the most powerful nations of the world understood that greenhouse gas

pollution was leading to dangerous climate heating that could destroy civilization, those leaders took decisive action to avert disaster.

In our universe, the reality was pretty much the exact opposite.

Because the conclusions of science were inconvenient to rich and powerful people who profit from fossil fuels, science took second place to profits. Most governments, led by the United States, did as close to nothing as possible to fight climate chaos. As President George H.W. Bush said at the Rio climate summit in 1992, "The American way of life is not up for negotiation. Period." In some cases, governments just continued to make the problem worse, by continuing to pay fat subsidies to oil drillers, gas frackers, and coal mining operations.

Against the gung-ho production of dirty energy and climate pollution by fossil fuel companies, the half-hearted efforts of business and government to cut some of that pollution clearly aren't working. Or they're just not working fast enough, which amounts to the same thing.

As Bill McKibben has written, when it comes to climate action, "winning slowly is the same as losing."

Science was never going to be enough to win on climate. Indeed, no kind of rational appeal on the climate would ever succeed against the wealth and power of the fossil fuel industry. Anybody who studied the abolition movement would have known that.

Frederick Douglass was right that it was going to take a fight to end slavery in the United States. In the face of the wealth and political power of slaveholding elites, logical or even emotional persuasion alone would not suffice, as abolitionists discovered by trial and error.

Early in their movement, New England abolitionists started out by trying to convince Southern slaveholders to give up their slaves voluntarily because slavery was morally wrong. They even printed up and mailed well-argued abolitionist tracts around the South to reach slaveholders directly with their heartfelt and logical appeals. But when Southern states responded with laws against sending "incendiary" literature likely to inspire a "servile insurrection" through the mails, abolitionists quickly learned that the genteel approach that they called "moral suasion" wasn't going to be enough.

With so much of their wealth invested in slaves, Southern planters were not open to arguments against the system. As historian James Brewer Stewart has written, abolitionists learned that to have any hope of success against slaveowners unwilling to give up their slaves, antislavery campaigners had to move from moral suasion to political confrontation.[2]

In the years before the Civil War, evolutionary biology was just getting started, and Darwin published *The Origin of Species* in 1859, just two years before the fall of Fort Sumter. But even without the benefits of modern knowledge about the overwhelming genetic similarity of humans of all races, mid-nineteenth century Americans had plenty of areas of knowledge telling them that slavery was wrong, from natural and social science to philosophy and theology. Progressive social thinkers continued to agree with the Enlightenment political thinkers known as the American Founding Fathers, who cited such sage lawmakers across history as Locke, Montesquieu, and the ancient Greeks to declare that all men were created equal.

Southern planters even accepted the truth of human equality as self-evident when they signed the Declaration of Independence, which was largely penned by ambivalent slaveholder Thomas Jefferson. However, within a few decades, as commodity crops on plantations worked by slaves grew more profitable, Enlightenment political philosophy started to become an inconvenient truth that conflicted with the way that Southern planters and their Northern allies in shipping, textiles manufacturing, and finance were making an increasingly lucrative living.

As the early nineteenth century wore on and King Cotton grew richer than King Midas, the white South developed alt-facts, bolstered by perverse versions of political science, theology, and even biology more in line with their bulging wallets. Proslavery Southern politicians who said that slavery was a "positive good" cited pseudo-science (like skull measurements) to argue that Black people were naturally inferior to whites. Keeping Black people in bondage was the best of all possible social orders because it protected and improved the inferior race, said spokesmen for slaveowners.

Whether inspired by the Southern planter elite's twisted use of science for propaganda or not, fossil fuel companies have waged a 40-year-long campaign to discredit climate science and spread their own alt-facts purporting to support such bogus stories as "the moral case for fossil fuels" or how more CO_2 in the atmosphere will "green the earth."

For years, climate experts tried to answer such claims one by one, fighting specious claims with science. This proved a frustrating exercise that yielded few benefits. On the plus side, after years of quoting science deniers in any story about climate change to provide "balance," eventually the news media stopped giving equal time with climatologists to renegade scientists, attorneys, and lobbyists bankrolled by oil and coal companies.

On the minus side, attacking their alt-facts hasn't seemed to stop science deniers from continuing to spread lies. Oklahoma Senator James Inhofe, the leading climate science denier in Congress, has repeated the same talking points about climate change being a "hoax" for three decades. It doesn't seem to have hurt his reputation among his constituents. Inhofe has continued to get reelected to the Senate the whole time. Yet, outside of oil patch states like Oklahoma, Inhofe's brand of open science denial has started to go out of fashion. Lately, as we'll see, pro-oil propagandists have come up with a more subtle approach to delay action on climate and against fossil fuels.

Meanwhile, back in antebellum America, once big cotton plantations started to generate fabulous amounts of money, abolitionists citing antislavery Bible verses or offering real facts about race and natural history didn't convince many nineteenth-century white people, especially in the South, to switch from proslavery to pro-emancipation. The South was unreachable for abolitionists. What did work was to convince a variety of types of people in the North who didn't directly benefit from trade with the South that slavery was either morally wrong or, more likely, that it gave an unfair economic advantage to big planters over white family farmers and tradespeople.

Self-interest was a less noble rationale for opposing slavery than a deep conviction that bondage was an evil inconsistent with republican government, Christian charity, and social progress. Fear of potential economic competition in new Western states from large slaveowners was a more compelling reason for many white Northerners to hate slavery and to vote against it when they finally had the chance to do so—by electing Abraham Lincoln president in 1860.

If they had understood how abolitionists had tried to change public opinion before the Civil War, climate activists could have avoided arguing with science deniers in the first place. Or, since it seemed that the news media until recently was bent on giving a voice to unqualified deniers on an equal time basis with scientists, that the climate movement should have been quicker to attack the very premise of balancing science with propaganda.

Freed from the distraction of arguing with paid liars, advocates for climate action could have focused instead on reaching out to diverse groups of the American public beyond the traditional environmental movement—from Black and brown people in cities to working white people in rural areas. Their message could have been that climate action wasn't just the right thing to do morally, but that it was also one of the best ways to protect the most vulnerable American communities. Fortunately, advocates of the Green New Deal have started focusing on how building out clean energy offers a path to prosperity beyond shuttered factories and beleaguered family farms. Promoting self-interest for unlikely allies is just one of the lessons of slavery abolition for oil abolition.

Historian James Brewer Stewart, the only American historian I could find who has made an analogy between fighting slavery and passing the Green New Deal, has concluded that the climate movement today has all the advantages of abolition without its weaknesses. Both movements have "vaulting aspirations." But while abolitionists had no policies beyond immediate emancipation, the climate movement has detailed plans with specific policies to transition to a clean economy.

> The Green New Deal exhibits all the vital strengths of the abolitionists, one of our history's most dynamic, disruptive, egalitarian, and ethically grounded social movements. At the same time, it is burdened by none of abolitionism's limitations even as it makes itself into a powerful hybrid mixture of a grassroots organization, political pressure group, guardian of the public good, and legislative insurgency.[3]

In another point to their credit that a historian can appreciate, climate campaigners have now started looking back not only at the geologic time of climate science but also at the historical time of political science.

The Sunrise Movement, 350.org, and other climate groups have begun mining American history for models of political movements so big that they could build the broad public consensus necessary to marshal the full resources of government and industry against an existential threat to the nation and the world. "Be it the mobilization around World War II or the New Deal itself, American history is rife with examples of the state setting out a bold goal and meeting it, leveraging the full force of the U.S. government to take on problems whose costs were simply unthinkable," writes climate journalist Kate Aronoff in a book published by Sunrise leaders.[4]

Looking back to the 1930s and 1940s will provide valuable lessons to implement big programs that link climate solutions to creating jobs for working Americans like the Green New Deal, once those programs are passed into law. But does twentieth-century history offer enough of a political precedent for getting government and business to accept something as large as the clean energy transition in the first place?

Why the Climate Movement Has Failed So Far

Along with the New Deal, the iconic political movements of the 1960s and 1970s—civil rights, anti-war, women's rights, and of course, the environmental movement—are offered as models for the strategies and tactics needed to put pressure on government and industry for serious climate action. In a very busy period, those movements achieved success against the forces of prejudice, ignorance, and pollution to pass watershed laws guaranteeing rights to political equality along with clean air and water. We need to repeat their successes now.

Initial moves by the Biden Administration on climate are promising. Yet, after more than 30 years of valiantly protesting, lobbying, and litigating, the climate movement has failed to pass any laws as big as the watershed civil rights and environmental laws of the 1960s and 1970s. Since then, even powerful legislation like the Clean Air and Clean Water Acts have been unevenly applied at best, requiring activists to go back to court in case after case to demand that regulators enforce their own rules.

The Clean Water Act alone was violated more than half a million times between 2004 and 2009, with no fines or any other punishment for most of the big polluters.[5]

Nonetheless, laws limiting pollution are necessary and they are a basis for lawsuits to require their enforcement. Yet even when laws succeed in reducing or cleaning up pollution, they can have unintended side effects that make things worse for people of color. As historian and ecological author Jenny Price explains:

> The environmental regulations we have allow us to do marginal improvements, but they don't address the root causes which are structural. Our economy, the engine of our society, is designed to ignore environmental and social costs. Sure, you can compare the black skies in Los Angeles from decades ago to today, and there has been improvement. But what we've mostly done is clean up pollution where most affluent people encounter it, and

then stash it and concentrate it in low-income areas. I think you have to ask whose quality of life is better? Are people who live in Cancer Alley, the strip between New Orleans and Baton Rouge where they have 150 petrochemical plants, do they think their quality of life is better than 50 years ago?[6]

Whatever gains have been made to clean up high profile bodies of water like Lake Erie or Chesapeake Bay or to clear the skies of Los Angeles, during the 30-year lifetime of the climate movement, the amounts of carbon dioxide and methane in the atmosphere haven't declined at all. Instead, climate pollution has gone up. And not just a little. In just the last three decades, humans in the Age of the Internet have spewed as much carbon dioxide, methane, nitrous oxide, and other greenhouse gases into the atmosphere as previous generations belched out from smokestacks and tailpipes in all of history until the late 1980s. In a tragic speedup, our ultra-globalized economy has gotten much better at doing damage to the climate. Sadly, our political leaders already knew better, but they let dirty industries that also knew better continue to damage the climate anyway.

Whose fault is it that in the exact 30-year period that we've all had the chance to know about the climate threat, the nations of the world and especially the United States have not only failed to make progress cutting climate pollution, but have made the problem twice as bad as before?

A variety of theories point fingers at different places. "Big Green" environmental organizations spent too much time lobbying bureaucrats in Washington and at the U.N. and not enough time building support among the broad public. Scientists scared the public too much or didn't scare them enough. Al Gore gave too many boring PowerPoints filled with ice-core data in charts and graphs. Climate change is too hard for ordinary people to understand or care about because it's the communications "problem from hell" that you can't see, smell, or feel, and that seems like it's always twenty years in the future or five thousand miles away. And my favorite theory, because it's so very wrong: Consumers were too selfish to care about the future and so they didn't do enough recycling and they kept taking showers that were too long.

Such notions are just distractions from the real culprit: lobbying and disinformation from the fossil fuel industry.

The facts belie all arguments that would blame the climate crisis on people who want to fight climate change. Dedicated leaders like Gore and ordinary activists alike have fought bravely and well, increasing public awareness over time, but the industry has killed big climate legislation and gutted international treaties—even when they enjoyed overwhelming public support. Brainy scientists have adapted quickly and skillfully to transform themselves into effective public communicators, trying both scary and hopeful appeals, and both have succeeded in different times and places. But lies widely spread across the news media by the fossil fuel industry have kept the public confused about solutions.

With painful experience gained in a decade of terrifying weather disasters, by now most Americans understand that climate is not just a problem for their grandchildren or for people on low-lying Pacific atolls. It's a problem for Americans living today on the Gulf Coast or in Arizona or in New York City. And ordinary people, as both citizens and consumers, have insisted in larger and larger numbers that politicians support clean energy and that companies sell green products. In response, politicians and businesses have stepped up, if not always and everywhere, at least in more and more places.

Though we all use dirty energy, ordinary Americans are not the ones to blame for our government's continued failure to deal with the crisis. The fossil fuel industry is to blame. In short, it has fought to keep doing business well into the future because oil, gas, and coal companies don't agree with carbon abolitionists that their time has passed. And the owners of those companies want to stay rich.

Mining coal might sound like an activity straight out of a Charles Dickens novel. And pumping oil and gas might seem better suited to a time when your average car buff yearned to drive a Trans Am rather than a Tesla. But, like a hairy chested version of Google or Apple, the average fossil fuel company has always prided itself not just on brawn, but also on brains. Over the last thirty years, with help from clever engineers, multinational petroleum corporations and family-run drillers alike got much better at using technology to extract oil, gas, and coal from deposits they would have formerly ignored as unprofitable.

Driven by crafty lobbyists, dirty energy companies also got much better at using propaganda to protect themselves from public anger over pollution. While spreading lies about climate science, fossil fuel companies also spent so much money on lobbying and political campaigns that they were able to convert the whole Republican Party into the political wing of the oil, gas, and coal industries. Dirty energy barons had less luck with (or spent less money on) the other side, but they still managed to discourage too many Democrats from trying very hard to cut down on fossil fuels.

After years of getting cheaper and more efficient, clean energy technology is ready to go. Solar panel prices have fallen by around 90 percent since the end of 2009, while wind turbine prices have fallen by almost 60 percent since 2010. In 2019, for the first time in history, the world added more solar and wind power, combined, than any other form of energy generation.[7] Meanwhile, advances in battery technology come along every month to make power storage more affordable. Cheaper, more powerful batteries will clear a larger place on the grid for renewable energy sources including solar and wind that generate power intermittently, that is, only when the sun shines or when the wind blows.

It's impressive—but not nearly enough and not fast enough. Most new energy may be clean energy, but plenty of legacy oil, gas, and coal-fired power plants are still running. The industry plans to keep those dirty plants running for decades to come, even if it makes the job of saving the climate that much harder. Dirty energy CEOs know that elected officials will keep giving them political cover as long as the campaign donations keep flowing.

And it's not just Republican politicians who are vulnerable to industry pressure. For example, during just nine months in 2020, the Democratic governor of dark blue California, Gavin Newsom, approved more than 1,500 new oil and gas wells, even walking back his own pledge to prohibit new fracking operations during the pandemic lockdown, under pressure—no surprise—from the state's oil producers.[8] More recently, even after pulling the plug on the Keystone XL pipeline, the Biden Administration bucked Indian tribes and environmental activists to give its support to construction of the Line 3 Pipeline in Minnesota that will transport carbon-intensive tar sands oil from Canada.[9]

Fossil fuel politics is the only reason it's taken so long to get any serious action on the national level on the climate crisis, even as a majority of the public has supported climate solutions. Because there's so much money behind fossil fuels, that's a logjam that we're unlikely to break with merely the tools of 1960s activism and sing-along acoustic guitar ballads. In the 1960s, racial prejudice was strong emotionally but not economically. If anything, big business stood to benefit as the end of Jim Crow would boost spending by Black consumers. Old timey segregationists like George Wallace and Sheriff Bull Connor had lots of hate, but very little cash, behind them. By contrast, ExxonMobil and the Koch group of companies, though somewhat weakened, are still some of the top contributors to political candidates.

As Chris Hayes has explained, under serious climate rules that would end dirty energy production in the near future, today's fossil fuel companies stand to lose about as much money from future potential sales of fossil fuel resources ($10 trillion, calculated in today's dollars) as Southern plantation owners were forced to give up when slavery was abolished after the Civil War. This suggests that the climate will need a movement now to fight Big Oil as big as abolitionists needed in the first half of the nineteenth century to fight what they called the "Slave Power," the unholy alliance of big Southern planters and business interests in both the United States and Europe who profited from slavery and counted on its expansion.

Fossil fuel companies and their allies are well organized and powerful enough to constitute a force in American politics and society that can only be thought of as the Oil Power. This "complex and highly-developed network of organizations built to block action on the life-threatening issue of climate change," encompasses an unholy alliance of dirty-energy and libertarian ideological interests. For more than thirty years, its members have successfully conspired to slow action on climate change and give fossil fuel companies more time to keep making profits, according to a report released in 2021 by the Climate Social Science Network and Brown University.[10]

This highly sophisticated denial machine brings to bear not just big money but crazy big money "to influence the public, media, and political arenas, with a goal to slow or stop ambitious action on climate" and consists of ten key types of people and groups, according to researchers behind the report.

The "Structure of Obstruction" begins with those who provide the funding for denial: large corporations like Exxon, Koch Industries, Dow Chemical, DuPont, Ford, and General Motors, as well as conservative philanthropies that allow wealthy families to donate money tax free, often in secret. That money goes to pseudo-authorities who give this denial machine credibility, including "contrarian" scientists who deny or downplay climate science, along with those including PR firms, "astroturf" (fake) activist groups, and free-market think tanks that generate climate misinformation and spread it to conservative media outlets like Fox News and bloggers who specialize in science denial.

Finally, Republican politicians fight public policy changes that would hurt fossil fuel companies. Together, this conspiracy of inaction has given fossil fuel companies political cover to keep polluting for decades, heating the climate and fouling the air and water of local communities.

By slowing down the clean energy revolution, the Oil Power remains a formidable foe to saving the climate. Fighting these counterrevolutionaries has required the climate movement to broaden its focus from cutting carbon to changing the whole economy. If we are to succeed at stabilizing the climate in time to avoid the worst, then the days are over of promoting standalone technical fixes to reduce greenhouse emissions from power plant smokestacks or to increase the gas mileage of a certain kind of car or truck by a few percentage points. The climate is now at such a precipice, scientists warn, that unless we make transformational changes across society to cut and reverse greenhouse gas pollution within the current decade, then humanity's future is unspeakably bleak.

Abolitionists felt the same way about slavery, freedom, and American society before the Civil War. Despite the name of their movement, abolitionists cared about many more issues than ending slavery. Theirs was a broader movement for freedom, fairness, and a meaningful life than history generally remembers. And while it started with opposing slavery, abolition expanded into trying to reform a "Hydra of Evils" that threatened American democracy. Slavery "had poisoned everything it touched," according to Wendell Phillips, explaining why he agitated not only for freeing slaves but also for universal suffrage and labor rights. "When we tore off the mask the same hideous features were behind it—a sneering and gibbering specter. This was America."[11]

Back in the 1730s, Quaker Benjamin Lay, perhaps America's first white abolitionist, combined antislavery with vegetarianism, refusing to wear or eat anything produced from the exploitation of either animals or slave labor. He published pamphlets against slavery, the prison system, and capital punishment. One of the innovators of in-your-face activism, Lay staged one-man protests at church meetings where he splattered the audience with fake blood or sat outside in the winter with his bare foot buried in the snow to remind passersby of the physical suffering of their enslaved brothers and sisters in Christ.[12]

In the years before the Civil War, abolitionists campaigned for a variety of reforms ranging from the deadly serious to the whimsical: socialism, votes for women, free love, celibacy, religious reform, extreme pacifism, and armed insurrection to free slaves across the South. A big abolition meeting held in Philadelphia in 1840 featured speakers for many related causes, including a vegetarian who "warned meat eaters that their diet was slowly causing their very bodies to convert into the flesh of cows, pigs, and geese. Another objected to this line of argument on the grounds that his body was not becoming a squash of a stinkweed."[13]

After they won their battle to abolish slavery, abolitionists went on to prominent positions in movements for Black education and civil rights, votes for women, and labor rights. They had built a movement that balanced a push for urgent, immediate action (to put out America's national "house on fire") with the patience required to win massive victories against powerful opponents. Abolition was able to endure for nearly a century, the time it took to accomplish its main goal of ending slavery. Even more impressive, their movement had a broad enough vision to connect freedom for slaves to economic fairness and civil rights for Black and white people alike.

In the same way, today we must achieve transformative change in the economy in this decade by fighting and winning a clean energy revolution. At the same time, we must plant the seeds for a movement long lasting enough to persevere another fifty or a hundred years to ensure that the revolution stays won.

We need to act fast to stop and reverse the pollution. But that's only the start. To keep cutting pollution, we'll need to convert climate care from an acute threat to be quickly met into a permanent lifestyle. And that will include redeeming our country's promises to ensure safety, freedom, and fairness for all Americans regardless of race or personal wealth.

That's why dealing with climate will take a much larger mass movement of Americans than we've seen so far, and a much deeper effort to transform many aspects of our society than we've seen in many decades. It will require the high intensity for immediate action seen in the civil rights, anti-war, and women's movements of the 1960s and 1970s, but with better staying power.

Locking in efforts to keep the climate safe and make the economy both clean and fair will also require the cool persistence to continue working for decades to come. And that will look more like the abolition movement, which managed to stick around long enough from its birth during the American Revolution, to keep attacking chattel slavery right through abolition's final victory at the end of the Civil War.

Learning from History and Hoping to Repeat It

The Green New Deal looks back to the 1930s New Deal and the total mobilization of industry that followed during World War II. This inspiring period offers a model of quick, powerful action with government as big as it needed to be to respond to an immediate international crisis.

But the New Deal, war, and post-war economic period only lasted for about 15 years, though historians have identified a "New Deal Consensus" favoring activist government that dominated politics until the election of Ronald Reagan as president in 1980. So, you can say that an extended New Deal lasted for 50 years. That's impressive. But the climate movement is already almost that old, and much work remains to be done. We will need an effort even longer lasting than the New Deal Consensus.

For a movement that endured long enough to transform our economy and political system over a longer period, we must look further back. The movement to abolish slavery, an evil that had plagued humanity for not merely centuries but for millennia, was the most successful movement ever because it was both the deepest and the most long lasting.

Chris Hayes has even written that the fight against climate change must become "the new abolitionism." To avoid heating the climate more than 1.5 degrees Celsius (2.7 degrees Fahrenheit), which scientists say is the maximum to avoid runaway global heating that will be beyond the ability to humans to slow or stop, more than 80 percent of remaining reserves of oil, gas, and coal must remain in the ground, unburned. Conservatively, assuming world oil prices at the low end of their range in recent years, Hayes estimates that those reserves are worth at least $10 trillion.

Convincing the powers that be from the governments of oil exporters like Saudi Arabia to the management of big oil companies like ExxonMobil to forgo so much wealth will be difficult to say the least.

31

As Hayes writes, imagine being the person who has to say this to an oil company CEO or board of directors: "That stuff you own, that property you're counting on and pricing into your stocks? You can't have it." That's the very definition of the world's hardest sales pitch.

Such a hard sales pitch, Hayes concludes, that it hasn't worked in more than 150 years. In fact, as we've seen from the failure of abolitionists' original approach of "moral suasion" to convince slaveholders to voluntarily free their slaves, asking nicely didn't work back then either. "The last time in American history that some powerful set of interests relinquished its claim on $10 trillion of wealth was in 1865—and then only after four years and more than 620,000 lives lost in the bloodiest, most horrific war we've ever fought."

Of course, Hayes is referring to enslaved people and to the Civil War, which he is correct to conclude was mostly about money. Despite putting out lofty rhetoric about violations of the Constitution and threats to states' rights, what Confederates who fired on Fort Sumter in April of 1861 were really fighting for was to preserve and even grow their largest source of wealth, the population of human beings they held in bondage. And as obscene as it sounds today to equate people with property—which was the issue over which the Civil War would be fought—just before the war, the enslaved people of the South had a market value of $10 trillion in today's dollars. Since that figure represented half of all wealth held by white Southerners, Southern planters decided that expanding slavery was lucrative enough to start a war over.[14]

Today, because it stands to lose so much money if the nations of the world get serious about preventing climate chaos and getting off fossil fuels, it's clear why Big Oil would also rather fight than switch.

Like abolitionists before the Civil War, climate activists today are calling for people who have bet on fossil fuels to forfeit a massive pile of future cash. The Slave Power wasn't willing to give up its wealth for moral reasons in 1860 and the Oil Power has shown so far that it's also not willing to lose out on all the future gains it had planned on just because its product is destroying America and the world—a fact that the industry has recently said that it accepts but has so far failed to act seriously upon.

It was a measure of the international abolition movement's success that slavery was outlawed in every nation on earth by the end of the twentieth century. Tragically, abolition's work to end bound labor also continues today in the fight to free more than 40 million people still held in illegal slavery around the world. So does abolition's successor, the civil rights movement, continue to fight for racial fairness and equality. But that doesn't cancel out the historic achievement of putting a legal end to a system of oppression reaching back to the dawn of civilization.

Abolition remains an unfinished quest. Yet, abolition's massive and nearly impossible victories of the past are well worth celebrating today. After centuries when people across the globe toiled in bondage, in the middle of the nineteenth century, the international antislavery movement succeeded in breaking the chains of millions of people in the British Empire, the United States, and across the globe.

If the climate and civil rights movements can achieve such victories in our lifetimes, we will be well on our way to a safe and just future. Applying abolition's lessons of success against the greatest of odds to today's fight to save civilization from climate hell with its built-in racism and threats to health, we will realize that we can afford nothing less than our own movement to abolish oil, gas, and coal. The oil abolition movement must also target the racism and wealth disparities that dirty energy both supports and relies on for its own support.

Naomi Klein urges us to look back for inspiration and ideas to our activist ancestors to give us strength. "The movement for the abolition of slavery in particular shows us that a transition as large as the one confronting us today has happened before—and indeed it is remembered as one of the greatest moments in human history."[15]

You Can't Talk about Climate by Itself Anymore

America today faces three intertwined crises: climate, race, and wealth inequality. At this moment of peril and promise, the three connect as they never have before.

Start with climate and its main driver, burning fossil fuels. Coal mines and coal-ash pits. Oil wells, petroleum refineries, and chemical factories. Fracking fields, natural gas pipelines, and methane compressor stations. Dirty energy facilities have fouled the air and water of local communities since the start of the Industrial Revolution. But they have hurt some people more than others.

People who live on the other side of the fence from such places, overwhelmingly low-income families and people of color, have gotten little benefit from companies producing, selling, and using fossil fuels that are sent to big cities and industrial customers. Jobs in coal or oil have always been dangerous and unhealthy for workers. But when unions were strong, at least working in a coal mine or on an offshore oil platform paid well and carried some pride. After decades of union busting and automation, that's less and less the case anymore. For example, the West Virginia coal industry shed 80 percent of workers from 1950 to the present. That wasn't because environmentalists were waging a "war on coal" as coal lobbyists claimed but because coal operators switched to automation to cut costs. Similar job cuts have decimated other fossil fuel sectors.[16]

Likewise, fossil fuel operations have always been bad neighbors for communities. When pipelines explode, when fracking chemicals leak into drinking water supplies, and when foul lakes of coal slurry burst their dams and flood into mountain valleys, local people are always the first to be hurt and too often, the last to be helped.

Suffering high rates of injury and death in accidents or asthma and cancer over lifetimes of breathing dirty air and drinking contaminated water, people in frontline communities have paid the highest price for Americans to enjoy cheap energy and for dirty energy CEOs and investors to enrich themselves. For a long time, those of us privileged enough to live in neighborhoods not blighted by nearby oil and gas pipelines, mining operations, fracking wells, tracks for coal trains, and gas compressor stations could look the other way. NIMBY was the codeword.

Martin Luther King Jr. famously said that "Injustice anywhere is injustice everywhere." Ever since, environmental justice activists have tried to convince Americans that pollution anywhere is pollution everywhere. Now, the climate crisis has made this point even harder to ignore.

Just ask families on the West Coast from San Diego to Portland whose suburban homes have been consumed by wildfires. Or you can ask people in high-rise condos in Miami whose cars float away in the high tides that now flood city streets or who fear more buildings will collapse from likely water damage to foundations as the Champlain Towers South collapsed in June 2021. Or those in the Southwest facing water shortages from continuous, severe droughts. Or in the Northwest experiencing high temperatures beyond anything they ever feared.

Under climate disruption, the poor and people of color will still suffer the most, even when hit by the same weird weather as their wealthier neighbors, who have more money to run air conditioning during a heat wave or evacuate before a storm hits and then rebuild or move away afterwards. To make matters worse, poorer people and people of color tend to live more in low-lying places and in the especially hot, dry, or wet places most vulnerable to weather disasters. And, in a triple whammy, those same people get less help to rebuild from government relief agencies.

When Hurricane Katrina hit New Orleans in August 2005, breaking levees and flooding the city, Black neighborhoods like the Lower Ninth Ward were most severely affected by the flood, whose waters took weeks to recede. Yet, Black homeowners received $8,000 less in aid than white homeowners, leaving a larger percentage of Black families with no place to live and with little help to rebuild or move someplace else.[17]

In frontline communities, the fossil fuel industry is adding insult to injury, and then more insult on top of yet more injury in an unending cycle. It's hard not to call that racist in outcome even if you can't prove any conscious intent to discriminate.

For years, many of us could pretend that pollution was a local problem for unlucky places. It's now easier to see that there's no NIMBY for global heating. Some cities will be hit harder than others. But there's no way to confine sea-level rise, hurricanes, floods, wildfires, and heatwaves to sacrifice zones. And no way to avoid massive migrations as, for example, coastal residents flee inland from increasingly untenable locations. Americans everywhere now have no choice but to pay attention to news about weird weather. And these days, weather news is more often bad news.

Consider a single day in August 2020 as I write these words. Researchers in Greenland reported that last year the country's ice sheet lost a million tons of ice per minute. Every second during 2019, enough of Greenland's ice sheet melted to fill seven Olympic-size swimming pools.

Meanwhile, in California, where another heat wave has led to another set of wildfires, authorities have told every single resident from the Mexican border to the Oregon line—not just those who live in areas usually vulnerable to fire, but all 40 million people in the state—to gas up their cars and then park them in their driveways facing out, to be ready to flee at a moment's notice.

Then, on the Gulf Coast, Hurricane Laura has started pummeling the coast of Louisiana, which happens to be home to much of America's chemical industry. High winds helped start a fire at a factory on the shore of Lake Charles that makes disinfectants and swimming pool additives, sending off smoke that could contain chlorine, nitric oxide, and other toxins.

Finally, in China, record rainfall is filling up the reservoir behind the Three Gorges Dam, reaching its highest level since it started holding water in 2003.

All that mayhem in just one day. What will happen tomorrow?

Slavery Didn't Abolish Itself and Oil Won't Either

A lot of the carnage can be blamed on the oil and gas industry, whose products, when used exactly as directed, pump climate heating gasses into the atmosphere. For decades, the industry was strong enough to fight off meaningful climate action by spreading lies about climate science and launching attacks on clean energy.

Even with sales down because of the Covid shutdown, the dirty energy industry still threw dollars at public disinformation campaigns. In 2020, the American Petroleum and Gas Association, an industry lobby bankrolled by oil and gas companies, was planning to spend more than $600,000 to convince Americans not to exchange their old gas appliances running on methane, the strongest greenhouse gas, for new electric ones that could operate on solar or wind power.[18]

This sort of thing is galling. But some experts think it may only be the death rattle of a toothless, senile industry limping towards the graveyard, helped along with a kick in the butt from the pandemic slowdown. Yea, how the mighty have fallen. As recently as 2011, ExxonMobil was the largest corporation on earth. But after the company's value declined from $525 billion in 2007 to about $180 billion in 2020, the Dow Jones Industrial Average dropped Exxon from its listing.[19]

The company had been a member of the blue-chip stock index for 92 years, longer than any other corporation. *Sierra* magazine ran its version of an obituary, a cover story called "The End of Oil Is Near," predicting that "the pandemic may send the petroleum industry to the grave."[20]

Looking back at history, many people today assume that market forces and evolving ideas of human rights would have ended slavery at some point, even without the Civil War. But history shows that would have been unlikely.

In the early days of American slavery, the Founding Fathers, even slaveholding ones like Jefferson and Patrick Henry, expected slavery to wither away on its own. Several Northern states had already begun to gradually free their enslaved people even before the Constitution was signed in 1789, and many white people assumed that Southern states would soon follow suit. In the early years after the Revolution, abolition societies were active in Virginia and other Southern states. But the invention of the cotton gin in 1793 changed the economics of Southern slavery, making it far more lucrative. And that was the beginning of the end of abolitionism in the cotton states.

The big money from growing cotton with slaves just kept getting bigger as new states in the South like Alabama, Mississippi, Louisiana, and Texas were added to the nation. By the early nineteenth century, slave state leaders were no longer talking about ending slavery or apologizing for what they called their peculiar institution. Instead, they began praising slavery as "a positive good" and hatching plans to extend slave agriculture to new territories out West and even to European colonies in the Caribbean that Southerners hoped to conquer.

Slavery was evil but it was also resilient, and it probably could have endured in the United States decades later than it did, as happened in other slave societies in the Western Hemisphere. The Spanish didn't abolish slavery in colonial Cuba until 1886, and Brazil didn't free its slaves until two years later, a full 23 years after the Thirteenth Amendment to the U.S. Constitution ended slavery in 1865.

Likewise, today, we should not count out Big Oil just yet. Ever since Edwin Drake poked a hole in the ground near Titusville, Pennsylvania and crude oil squirted out in 1859, the oil and gas industries have successfully weathered boom-and-bust cycles that would have sent other industries straight to bankruptcy court. Coal has endured even longer and survived even more market ups and downs. From the Depression of 1815, the Panic of 1873 and the "Long Depression" of the 1870s that followed, to the Great Depression of the 1930s, and the Great Recession of 2008, fossil fuel producers have seen demand and prices plummet but have lived to see them rebound afterwards.

If any industry knows how to ride the waves of a volatile market, it's fossil fuels. If Wall Street and big banks keep the money flowing for ExxonMobil, Shell, Texaco, and their friends to keep drilling wells in ever more remote and hard-to-reach places, then dirty energy will survive today's turmoil. Smaller frackers and pipeline operators may go bust. But the bigger ones will just buy up their assets at liquidation sales and keep on mining, drilling, and dumping as long as they can.

Peak oil? We've heard that before, but somehow, the industry has always managed to bounce back with new supplies or at least new financing schemes to keep selling fuel cheaply enough to discourage competition from alternatives. But this time it really could be different.

Aside from a heightened public concern over climate, the big change this time around is that renewable energy has finally reached maturity. In the last few years, solar and wind crossed the line from experimental and expensive "alternative" energy to shovel-ready and affordable "least-cost" power, to use a term favored by electric utility regulators. Demand for fossil fuels may have peaked in the American market. And in September 2020 most executives of oil and gas companies in Texas said that U.S. oil production had finally peaked also, according to a survey by the Federal Reserve Bank in Dallas.[21]

A prospect for future growth was one way that slavery was different before the Civil War than oil is today. Slavery was at the zenith of its profitability in 1860, on the eve of the Civil War, judging by its main product, cotton. American cotton production soared from 156,000 bales in 1800 to more than 4,000,000 bales in 1860. And the enslaved population of the United States was also at its peak in 1860, growing from 700,000 people in 1790 to nearly 4 million souls in 1860.[22] Without forced abolition after the Civil War, slave-grown cotton could have kept expanding for years more.

By contrast, fossil fuel companies have taken such a beating in the last few years and especially during Covid that they are clearly an industry on the way down. The issue now it not whether fossil fuels will survive—they won't—but how quickly they'll go away. Cutting climate pollution is urgent. But the end of oil, gas, and coal production could take decades and, despite assurances that they want to be "part of the solution" on climate, fossil fuel companies have shown no willingness to hurry up the process of ending fossil fuels. Quite the opposite.

In 2018, Exxon CEO Darren Woods told investors that ExxonMobil was planning to invest $30 billion every year to build facilities using the latest technology that could produce large amounts of oil and gas around the world for decades to come regardless of price fluctuations in energy markets. Woods promised investors that the company would exploit "the richest set of opportunities" it had seen in years, including shale oil in the Permian Basin in southwestern Texas and New Mexico, offshore oil in waters belonging to Guyana and

Brazil, and liquefied natural gas in Mozambique and Papua New Guinea. The company is secretive about its plans, but in the middle of the Covid shutdown in July 2020, Exxon said that projects put on hold during the pandemic would be restarted again afterwards.[23]

Meanwhile, surely Woods believed that if oily checks kept flowing to political campaigns, then government subsidies would continue to keep gas and oil flowing too.

Take the example of Covid relief funding. Before the pandemic hit, many petroleum companies had been cutting benefits for workers and eliminating jobs for years as their businesses declined under low oil and gas prices. But that didn't stop the Trump administration from using the pandemic as an excuse to gut environmental controls while awarding the oil industry billions of dollars in relief intended specifically to preserve jobs. "The oil industry has been shedding jobs for years. There is no evidence that if you roll back regulations, invest more in fossil fuel companies, that it will result in more jobs," says economist Belinda Archibong.[24]

But creating jobs wasn't really the point. The purpose of government handouts to dirty energy companies is to make big political donors among oil CEOs and investors happy. This is one reason why America's rules on campaign contributions must be reformed.

In the meantime, even as demand for their products declines, fossil fuel companies will be able to live on government assistance. Big Oil has grown fat and happy on taxpayer subsidies for more than a century, and they wrote the book for all American industry on how to keep the government money pipeline flowing.

Indeed, the fateful Supreme Court decision that gutted campaign finance laws and opened a whole new level for big money to buy politicians in 2010, *Citizens United*, was masterminded by oil companies hiding behind a front group. The decision "allowed fossil fuel political power to effectively capture Republican elected officials nationwide," according to a U.S. Senate report from 2020.[25]

"Prior to *Citizens United*, there had been a long history of bipartisanship on climate," dating back to the early 2000s. Republican John McCain even ran for president with climate action in his platform. But after the *Citizens United* decision, GOP officials fled from climate as an issue, fearing retribution from big fossil fuel political donors like the Koch brothers.

Why did the dirty energy industry target Republicans in particular, instead of just spreading their largesse—and making their threats to support a candidate's opponent unless the candidate toes the industry line—across the most promising candidates of both parties? "Fossil fuel executives realized that they only needed to keep one party in line, especially given Senate procedural rules that make it difficult to pass legislation without at least some bipartisan support," says the Senate report. "They made the strategic decision to target Republican officials. As the party traditionally more aligned with business interests—and more dependent on business interests for political funding—they were also an easier target."[26]

This strategy paid off. The politics of dirty energy money after *Citizens United* bought the industry another ten years in which to pollute without having to pay the costs. But oil industry planners know that they can't enjoy this free pass much longer. With each passing year, Americans get more and more frustrated at the lack of action on climate while at the same time it becomes easier and cheaper to transition to clean energy. The industry predicts that serious political action on climate will eventually force cuts in government subsidies, letting the market finally decide that fossil fuels are too expensive to run cars and heat homes.

Yet, oil companies are not ready to fade away into the sunset just yet. Instead, while fighting a desperate rearguard action to save as many subsidies as they can for as long as possible, the industry plans to reinvent itself. Oil company business strategists are nothing if not imaginative, and they are now touting dozens of ways to let them keep producing oil and gas while allegedly supporting climate solutions.

One of their most dangerous ideas is to turn petroleum into more plastic in more places. The industry has spent more than $200 million to build plastics plants over the last decade. Big Oil lobbyists are also pushing to repeal bans on single-use plastics like shopping bags and straws across the U.S. and around the world. If you believe oil company business plans, the future in processing petroleum into throwaway plastic convenience goods is looking bright.[27]

Given how resilient the fossil fuel industry has always been, it should be clear that the valves on oil and gas pipelines will only shut when a much bigger push for climate action comes along. That push must include derailing the industry's planned "plastic renaissance" along with many other bad ideas they're touting to keep pumping oil and gas that they claim are guilt free. Deceptive schemes range from carbon offsets to carbon capture and sequestration. If we fall for any of these tricks, or if we wait for market forces to make fossil fuel companies unprofitable on their own, it could take half a century or more to abolish oil. And that would be too late for the climate and too late for people. Those people being us, of course.

"I Can't Breathe": Racial Justice, Climate Justice

Unnatural disasters like heat waves, coastal storms, floods, and droughts that are caused or made worse by climate heating usually hurt poor people and people of color more than wealthier and whiter folks, and so do fast-spreading infections like Covid.

During the pandemic, Black and brown Americans were more likely to be exposed to coronavirus because they worked in service jobs. Meanwhile, Black and brown people have always been more likely to suffer from health problems like asthma, often connected to pollution. And they're more likely to lack access to adequate healthcare.

Clearly, the work of making America both free and fair for all citizens didn't end with the Thirteenth Amendment in 1865. That amendment abolished slavery or involuntary servitude in the United States, "except as a punishment for crime whereof the party shall have been duly convicted." This turned out to be a loophole big enough to float a supertanker through. Right after the end of slavery, Southern states passed Black Codes to ensnare freed people in convictions for minor crimes such as vagrancy that could lead to a stint at forced labor or convict leasing programs, sometimes working on plantations for their old masters.

In the 150 years since, Black men have been disproportionately targeted by the criminal justice system, often winding up in prison labor programs working for private companies at pennies an hour. It's no wonder that a movement has arisen to reform convict labor and even to abolish prisons in the name of racial equity.[28]

After the Civil War, to their credit, even after the partial but still incredible victory represented by the Thirteenth Amendment, abolitionists knew that their work had to go on. Americans of conscience have continued to work to secure civil rights for Black Americans from post-Civil War Reconstruction up to the present day. Today's Black Lives Matter movement continues the fight for racial justice that eighteenth- and nineteenth-century abolitionists started.

Environmental justice advocates like Robert Bullard have explained how fossil fuels have for years disproportionately affected people of color. After decades, now in his seventies, Bullard is still demanding equity: "[It's] time for whites to stop dumping their pollution on people of color," he wrote.

> Pollution is taking a heavy toll on the health of African Americans and other people of color. For example, a 2017 Harvard University study found African Americans are nearly three times more likely to die from exposure to airborne pollutants than other Americans. According to the CDC, African Americans are almost three times more likely than whites to die from asthma related causes; African American children are four times more likely to be admitted to the hospital for asthma, as compared to non-Hispanic white children; African American children have an asthma death rate ten times that of non-Hispanic white children. Dismantling institutional racism would go a long way in closing environmental health disparities in the United States. It's time for this immoral, unjust, and illegal pollution dumping to end.[29]

Civil rights leaders including Rev. William Barber have joined calls for environmental and climate justice. Naomi Klein, Bill McKibben, and other white climate leaders have also started talking about climate justice. "The one iron law of climate change is, the less you did to cause it, the sooner you suffer," says McKibben.[30] The summer of 2020 made George Floyd's cry for help while being choked to death by Minneapolis policeman Derek Chauvin "I can't breathe" into a rallying cry for clean air also.

As McKibben has written, "having a racist and violent police force in your neighborhood is a lot like having a coal-fired power plant in your neighborhood. And having both? And maybe some smoke pouring in from a nearby wildfire? African Americans are three times as likely to die from asthma as the rest of the population. 'I Can't Breathe' is the daily condition of too many people in this country. One way or another, there are a lot of knees on a lot of necks."[31]

Not surprisingly, Black and brown Americans are much more likely to prioritize climate and much more willing to join groups for climate action than white Americans.[32]

On the other side, polling has shown that white voters who score highest for racial resentment also are the biggest climate science deniers. Businesses or consumers acting on their own can't make anywhere near enough impact to save the climate. Serious climate action will require a total government mobilization of the economy like the Green New Deal. To stop serious government action, fossil fuel companies like Exxon and Koch Industries have supported both climate science denial and a free-market and "small-government" ideology claiming that government spending overwhelmingly favors people of color and immigrants at the expense of hard-working white people. One need only consider a century of government subsidies to fossil fuel companies, owned mostly by rich white men, to expose this story as a lie.

As public opinion analyst Ian Haney López writes, "The basic strategy of the Right—including major polluters like the Koch industries—is to fund racist dog whistling as a means to break popular confidence in government...To be clear, [the climate] movement must tackle racism. It cannot succeed simply by focusing on concerns like the economy or the environment alone."[33]

Elections Are the Start, not the End, of the Fight

It took the election of Abraham Lincoln in 1860 to finally flip the U.S. federal government from proslavery to antislavery. But this change in attitude was limited. Whatever he might have felt privately, as president, Lincoln was elected only to stop the spread of slavery to new territories in the West, not to abolish slavery in Southern states where it was already allowed.

After his election, Lincoln tried to calm Southern leaders by promising that he would not try to end slavery in their states. But that reassurance wasn't good enough for the planter elite. They were counting on expanding slavery to new territories. Through bad agricultural practices, they had exhausted their soil and needed to fresh land to keep cotton production high.

Also, to keep antislavery opinion from swaying the federal government against them as the nation expanded westward, Southern leaders insisted on keeping parity in the Senate between slave states and free states. Planters weren't prepared to settle for being outnumbered by the growing population of the North; denied room to expand slavery as the nation moved West and see slavery hemmed into a restricted area in the South; and ultimately, outvoted in a Northern-dominated Congress that could potentially decide to abolish slavery.

There was no such thing as steady-state slavery. The political, economic, and social system of slavery needed to keep growing or else it would ultimately die. That's why when Lincoln was elected, slave state leaders decided that it was time to secede from the Union.

Because Southern planters couldn't make the national government serve their regional interest anymore, they decided to start their own country, one that would never put any limits on the growth of slavery. The Confederacy was explicitly not dedicated to the proposition that all men were created equal. Instead, the new nation was conceived in unfreedom and dedicated to the exact opposite principle, that the white race should always rule, that Black people should always be slaves, and that slavery should always be able to expand.

In the fall of 2020, the election of Joe Biden was a welcome development for climate action. And it's unlikely that the Oil Power will try to secede from the Union, or any modern equivalent, just because a Democratic administration has finally pledged to take climate more seriously. After all, fossil fuel companies still have plenty of influence in Congress, on the Supreme Court, and pretty much everywhere else in Washington.

The climate movement can't just rely on having the right president or even the right Congress in Washington to fix everything. Yes, Democrats take each of today's crises more seriously than Republicans do. Democratic voters say that climate change is a top issue, surpassed only by racial inequality and healthcare. Among Republicans, climate change now ranks as the...wait for it...least important issue. You can blame that on *Citizens United*: after GOP officials stopped working in good faith on climate solutions in 2010 and started spreading denial of climate science, their voters followed their leaders' example.

Given Big Oil's control over the Republican party, electing Democrats will be necessary to enact the largest-scale climate programs like the Green New Deal. When Republicans take back control of Congress or the White House in the future, they will block anything that could hurt their big donors in dirty energy industries. That will mean undoing as many of Biden's climate programs as possible. This is a stomach-churning political seesaw that Americans have gotten used to riding ever since climate entered the public conversation in the late 1980s. When Democrats gain power, they enact climate solutions. Afterwards, when Republicans gain power, they undo those climate solutions. Rinse and repeat. It's no wonder that the rest of the world has learned to be skeptical of any claims that Americans are ready to "lead" on climate solutions.

That's why it's so urgent for people who care about climate to reduce the power of fossil fuel companies when we have a chance. Because once Republicans are back in power, those companies will again have enough influence in Washington to get their way for the term of Congress or the term of the next president. We must remove the "social license" of fossil fuel companies to operate and break this cycle now or else the climate will be fried. Once Big Oil loses all credibility with the public, then Republican politicians will no longer feel the pressure from fossil fuel lobbyists to block any serious action to save the climate.

To put it bluntly: to save the climate and keep it saved, oil companies must become as despised as slaveowners. Spreading the truth about their despicable campaign to lie to the public for years about the true danger of their pollution will be crucial to reducing their political power.

Even if it were possible, electing Democrats forever would still not be enough. You don't need to go back to the Civil War or even the New Deal to see that a president, even a sympathetic one, can only get so much done, even with a majority in Congress. When Barack Obama was elected in 2008, he promised to get serious about racial equity, climate solutions, and of course, healthcare. And for the first two years of his administration, Obama had a Congress controlled by Democrats to help him.

Well, one out of three isn't bad—in the face of determined Republican opposition, America did get the Affordable Care Act in 2013. Except that getting Obamacare wasn't the same as keeping it. While Obama was president, Republicans in Congress managed to stage an even 100 votes to repeal or water down the healthcare law.[34]

After Donald Trump was elected in 2016, Republicans finally succeeded in significantly watering down, if not gutting, Obamacare. Then, unaccountably, Trump cut programs meant to identify outbreaks of disease around the world and predict epidemics, leaving Americans much more vulnerable to the coronavirus.

If you're playing to beat the climate crisis, the game didn't end with the last election, and it won't end with the next one. Someday, Republicans will certainly become more reasonable about climate science and will be more open to bigger climate solutions, as they were in the past. Voters can put pressure on GOP candidates to get serious about climate, as they have done successfully in Florida, where ignoring sea-level rise is nearly impossible for any politician, no matter how much pressure they get from party bosses and big donors. And a small number of Republican leaders have already started to push from inside the GOP to make the party accept climate science and propose its own climate solutions in line with free-market economics and conservative values. We will meet one of these Republicans, former South Carolina Congressman Bob Inglis, in the gallery of climate heroes at the end of this book.

In the White House, even the best president needs to be pushed by activists. He may need the political cover that activists provide. Consider what happened after FDR was elected in 1932. Now that he would be in the White House, activists asked him to follow through on campaign promises to put Americans back to work during the Depression. "I agree with you, I want to do it," the new president replied. "Now make me do it."

Before the Civil War, abolitionists considered Northern politicians to be weak on antislavery and too prone to compromise with Southern leaders to avoid conflict. Leading abolitionists often urged Northern states to secede from a Union with Southern slave states long before the Confederates beat them to it. And when Southern leaders did threaten secession at various points in the early nineteenth century, abolitionists usually urged the federal government to just let them go.

But after the fall of Fort Sumter, abolitionists did an about-face, denouncing Southern secession and urging on the Northern war effort, while pushing Lincoln to make the war about abolition. Their support was crucial in giving Lincoln political cover to issue the Emancipation Proclamation on January 1, 1863, against opposition from much of the Northern public, from many politicians in his own Republican Party, and even from members of his own cabinet.

Black abolitionists also lobbied Lincoln for months to accept Black men into the U.S. Army and Navy to earn respect that would help white Americans recognize their equality. "Once let the Black man get upon his person the brass letters U.S., let him get an eagle on his button, and a musket on his shoulder and bullets in his pockets, and there is no power on earth which can deny that he has earned the right of citizenship in the United States," said Frederick Douglass.[35]

Lincoln was afraid that arming Black men would scare leaders of border states like Kentucky and Missouri, that still had slavery but had not joined the Confederacy, into leaving the Union. But after a year and a half of fighting, with the Emancipation Proclamation, Lincoln welcomed Black men into the fight. By the end of the war, about 180,000 Black men had enlisted in the army and another 19,000 in the navy. A high proportion of Black men paid the ultimate price to save the country and end slavery, with much higher death rates than their white counterparts because Southern troops targeted them for savage treatment.

Black women, who could not formally enlist as soldiers or sailors, contributed by serving as nurses, spies, and scouts. The most famous was Harriet Tubman, who scouted for the Second South Carolina Volunteers, a regiment of Black soldiers. But Tubman went beyond the roles allowed to women during the war and became the first woman to lead U.S. Army troops into battle at the raid on Combahee Ferry in 1863.

It took pressure from abolitionists to push Lincoln to act against opinion in border states. Abolitionists helped convince the Northern public that help from Black fighters was necessary to win the war, which was dragging on far longer than most people had expected. This change in public opinion gave Lincoln the political cover to implement Black enlistment.

Despite promising initial moves on climate, some climate activists see Joe Biden in much the same way that abolitionists viewed Lincoln: a well-intentioned centrist who is liable to water down any serious climate action in compromises with his old colleagues in Congress. Biden has said that he supports a transition to clean energy, though not as fast as the Green New Deal calls for. But in exchange for the support of Alexandria Ocasio-Cortez and other GND sponsors, Biden agreed to adopt many of their proposals.

Ocasio-Cortez promised to hold Biden to his promise while pushing him to do more. "We're not going to forget about that agreement, for the sake of an election, are we? What we're gonna do is that we're going to organize and demand that this administration—which I believe is decent, and kind and honorable—keep their promise," she said in a speech outside the headquarters of the Democratic National Committee just after Biden was elected.[36]

The Civil War historian Eric Foner has compared Ocasio-Cortez and other progressives to abolitionists. Just as progressives today try to pull a centrist president to the left to take more radical action on climate solutions, so abolitionists worked to pull the centrist president during the Civil War, Abraham Lincoln, further to the left towards more radical action against slavery. "In times of crisis," Foner told the New Yorker, "people with a clear ideological analysis come to the fore."[37]

Today, Ocasio-Cortez's supporters in the Sunrise Movement predict that we'll need 11 million or so Americans as "active and sustained" campaigners on the frontlines to attend strikes, marches, and political events to help generate the political will for serious climate solutions like the Green New Deal. Then, we'll also need millions more sympathetic supporters at home to vote for political candidates who will get with the program. The history of abolition shows that reform can go very slowly for decades and then, suddenly, once the movement reaches a tipping point in public support, big changes can happen quickly. And all it takes is for a committed and vocal minority to recruit about 25 percent of the public to their view to reach that key tipping point, according to a study done in 2018.[38]

After the climate justice movement makes elected leaders take the first serious action on climate, equity, and health, the movement will need to keep making them take more action. Getting off oil will take far longer than one presidential administration or a couple sessions of Congress. We'll need to keep the pressure on over many cycles of the federal government into the future. And that will mean building a movement with the kind of staying power found in few movements of the post-war period.

That's why we must go all the way back before World War II, before the Great Depression, and back into the nineteenth century to examine what happened in abolition. It took more than the four years of the Civil War to end slavery in the United States. It took 90 years from when the first abolition society in the United States was founded in 1775 until the passage of the Thirteenth Amendment ending slavery in December of 1865. And even then, the work of freedom and equality for Black Americans still wasn't done. It's still not done today. So that's a movement that's nearly 250 years long and counting.

Will it take 250 years or even 90 years to get control of the climate crisis? Scientists warn that we don't have that much time.

Really, we only have until about 2030 to stop the worst effects of atmospheric heating before big feedback loops cause heating that feeds on itself beyond human control. But even if we are successful in making changes in the economy big enough to count in the coming decade, there will be many more decades of work to undo the damage to the atmosphere, the oceans, forests, and other natural places—not to mention the cities, suburbs, towns, and farms that are part of any complete picture of the environment.

Big Tent Abolition Movements, Then and Now

The abolition movement was not for everybody. Big Southern planters and most ordinary Southern white people never got on board, and the planter elite's unwillingness to accept any limits on the expansion of slavery became the cause of the Civil War.

But outside of the white South, the antislavery movement was a big enough tent to shelter all kinds of Americans, a fusion coalition of Black and white people. Some Americans opposed slavery for moral reasons and were inspired to join a crusade for a noble cause. Others didn't care much about the fate of enslaved people one way or the other; but they hated the arrogance of slaveowners and feared economic competition from big plantations.

For millions of enslaved and free Black Americans, abolition was about life itself. Leading Black abolitionists like Frederick Douglass and Harriet Tubman made abolition a moral crusade to expand the nation's promise of freedom and equality to all Americans.

Then, a whole spectrum of white people opposed slavery for their own reasons.

On the far left, activists like William Lloyd Garrison and members of Congress like Massachusetts Senator Charles Sumner and Pennsylvania Representative Thaddeus Stevens denounced slavery as a moral abomination. In the center, pragmatic politicians from Abraham Lincoln on down sought to first stop slavery's spread and then, after Confederates fired on Fort Sumter, to eliminate slavery altogether as a cause of sectional conflict. And on the right, small farmers and tradesmen in the so called "Free Soil" movement sought to exclude slavery from new territories in the West to keep large planters from dominating those areas both politically and economically.

Sadly, many Free Soilers wanted to keep slavery out of new states because they didn't want economic competition from Black workers, whether free or enslaved. This is where most antislavery Northerners differed from abolitionists, who wanted to both emancipate the enslaved and grant them full citizenship, civil rights, and perhaps even social equality. As historian Eric Foner explains, "it was perfectly possible to be genuinely antislavery and deeply racist at the same time. In fact, probably most Northerners…were exactly in that boat."[39]

The Northern antislavery movement was not 100 percent morally pure. But it didn't need purity to achieve success. With a few white Southerners of conscience thrown in, this coalition of unlikely allies disagreed about many things. But on one crucial thing they were united: they had to break the hold of the Southern planter elite and their allies over the U.S. government.

That same unity of purpose will lead today's climate movement to victory against one of the world's greatest and most oppressive combinations of wealth since the Slave Power—fossil fuel corporations and their allies. The modern movement to abolish oil will build on today's movements for climate solutions, environmental justice, and social equity, a diverse coalition of young and old, urban and rural, many Democrats and some Republicans, immigrants and native-born people, LGBTQ+ and straight people, people of faith and atheists, and even people of conscience from inside the oil, gas, and coal industries.

The next chapter will examine why crises that threaten America's very existence seem to come in cycles whose limits are marked by epic crises that threaten to destroy our society.

There was nothing inevitable about ending slavery in the mid-nineteenth century—and there was nothing inevitable about preserving the American experiment in democratic government either. If things had gone the wrong way in the years leading up to the Civil War and in the outcome of the war itself, slavery could have continued and even expanded for a decade or more in the U.S., ready to spread from the American South to Caribbean and Central America.

Likewise, there's nothing inevitable about the climate crisis and the future of humanity. It could go either way. Looking at how Americans helped the moral arc of history bend towards justice at a time of great peril in the past will help us bend that same arc yet again in our own time of peril today.

2

History on Our Side

There is a mysterious cycle in human events. To some generations much is given. Of other generations much is expected. This generation has a rendezvous with destiny.

—Franklin Delano Roosevelt

Martin Luther King Jr. said that "the arc of the moral universe is long, but it bends towards justice." The story of his life shows that King didn't just wait around for the unstoppable march of time to automatically right the wrongs of the past. Instead, he joined millions of other Americans to fight for racial equality, a fair economy, and an end to the Vietnam War. When an assassin's bullet ended his life at a motel in Memphis in 1968, King ended up making the ultimate sacrifice for his cause.

King knew that big progress on civil rights and economic fairness was likely given trends in American society in the booming economy after World War II. But he also knew that success was not inevitable.

In the same way, King's ancestor in activism from a century earlier, the abolitionist Frederick Douglass, also knew that the trend of American history was tending towards freedom and equality, even if the journey passed through many injustices along the way.

But Douglass didn't expect slavery to just fade away by itself as Americans got more enlightened politically or the Industrial Revolution helped machinery replace human muscle power. If nothing was done by the government to end slavery, Douglass wrote in his newspaper the *North Star,* then as long as American cotton exports continued to grow, the number of people held in bondage in the United States would keep growing too.[1]

He knew that permanently ending slavery in America would take a political fight—and, perhaps, a military fight too. Before the Civil War, the alliance of Southern planters and their allies who made quick money off slavery had too much control over state governments in the South and over the federal government in Washington, DC to allow for slavery to be abolished in the South in an easy and orderly way, as it already had been abolished in six Northern states and throughout the British Empire. "Power concedes nothing without a demand. It never has and it never will," Douglass warned. It turned out that he was right.

Looking back from today, it seems natural that an institution so wrong in every way as slavery would have come to a well-deserved end around the time when it did, even without the abolition movement and the Civil War. But history shows that neither the forces of the marketplace nor public disapproval would have been enough by themselves to replace enslaved labor with machinery.

Up until the very end in 1860, slave-powered commodity agriculture was growing in profitability. At the time, those who profited off the sweat of bondsmen's brows expected to get even richer in decades to come from expanding slavery to new areas out West and possibly even by conquering Caribbean and Latin American nations. "The introduction of more negroes, and the extension of slave territory, are new doctrines with us," wrote proslavery Virginian George Fitzhugh just before the outbreak of the Civil War. "Give the North a little time, and she will eagerly adopt them."[2]

Fortunately, Fitzhugh was wrong. But it took the bloodiest war in American history to prove it. The economy didn't end slavery on its own, and contrary to popular belief, the economics of slavery were unlikely to make bound labor disappear on its own in a region with no source of wealth able to compare to slave-powered commodity agriculture, an economy where "cotton is king; and rice, sugar, Indian corn, wheat, and tobacco, are his chief ministers," according to Fitzhugh.[3]

Ending slavery required a combination of economic changes plus a huge political movement dedicated to confrontation that broke up the United States.

Today, economics alone have not been sufficient to end fossil fuels even though the economics of clean energy have never been better. For solar power alone, the installation cost has declined three hundred times over the last four decades from $76 a watt in 1977 to 25 cents a watt in 2017.[4]

And please keep in mind that the most obvious financial benefit of clean energy like solar and wind is not that it's clean, but that its fuel cost is zero. The sun and wind are free.

A traditional plant to generate electric power from heat must be fed a steady supply of fuel to burn, whether coal or natural gas or uranium for a nuclear plant. Once built, operations and maintenance costs for solar and wind are low compared to what it takes to run a thermal power plant. A nuclear plant is especially expensive to run. Highly skilled engineers must operate under complex safety protocols that conclude in the careful storage or disposal of radioactive waste. And when an accident happens at a nuclear, coal, or gas power plant, containment and cleanup costs range from merely exorbitant to catastrophic. By contrast, as the old social media meme goes, "When there's a huge solar energy spill, it's called a 'nice day.'"

Not just cleaner, but cheaper too. In every way, once equipment is installed, it's cheaper to go with solar and wind. But that's not been enough to replace fossil fuels with renewables as fast as companies could develop solar and wind power.

Political opposition from oil, gas, and coal companies has stood in the way of clean energy. So, we cannot rely on the operation of supply and demand alone to make the economy clean, especially since the energy market is distorted by huge subsidies for dirty energy. To give America the transition to climate solutions that we need, the energy market requires a big push from politics. For any truly transformational change, that's how it's always worked in the past.

Case Study of a Battle Nearly Lost

Without the right political push, and a bit of luck, changes in society that seem inevitable to us today might not ever have come about when they did. History doesn't always go forward towards better technology in the economy or more equality and fairness in politics. Progress is not an unstoppable force, mostly because different groups of people disagree about what qualifies as progress.

The presidency of Donald Trump ably demonstrated how special interests can turn back the clock on issues ranging from social equity to climate. More importantly, even when progressive changes are successfully made, sometimes they just barely squeak through—which shows that if activists didn't fight as hard or didn't get as lucky, the change would not have happened when it did.

Take as an example one of the biggest political changes of the early twentieth century, women's suffrage. Today, votes for women in the early twentieth century might seem like the natural outgrowth of an industrializing economy with corresponding changes in social attitudes: as the industrial sector grew in the Gilded Age, factories demanded more and more women to work outside of the home or off the farm. Yet, after it was passed by Congress following World War I, the Nineteenth Amendment giving women the vote nationwide almost failed to win approval by three fourths of the state legislatures required to add it to the U.S. Constitution.

By 1919, women already enjoyed some voting rights in 36 states. They were only barred from all elections in 12 states. Some suffragists thought that the issue could be handled entirely on the state level and that a federal amendment giving the vote to women across the United States was not necessary. Meanwhile, opposition to national women's suffrage was intense. Many opponents of suffrage were women themselves and one prominent group published a widely read pamphlet claiming that "90 percent of women either do not want [the vote] or do not care."[5]

Interestingly, along with reasons why women should not vote, this bizarre pamphlet also offered household tips including "You do not need a ballot to clean out your sink spout" and "There is...no method known by which mud-stained reputation may be cleaned after bitter political campaigns."

After nearly all other likely states rejected the amendment, only Tennessee remained in 1920 as perhaps the last chance to become the 36th and final state needed to ratify the amendment. The measure easily passed the state senate, but the suffrage amendment stalled in the Tennessee House of Representatives. When the state legislature finally held its vote on a muggy August day in Nashville, the amendment passed by a nail-biting margin of 49-47 only after state representative Harry Burn, the youngest member of the state assembly, switched his vote from no to yes to please his mother.

Arriving at the state capitol building that morning wearing the red rose of anti-ratification on his lapel, Burn was handed a note from his mother. In the letter, Mrs. Burn praised suffrage leader Carrie Chapman Catt and asked her son to "be a good boy and help Mrs. Catt put the 'rat' in ratification." Without Harry Burn's mother, the vote would have been tied and the amendment would have lost in Tennessee by default.

Given that other states didn't ratify the Nineteenth Amendment until as late as the 1970s (Louisiana and North Carolina) and 1980s (Mississippi), if suffragists lost in Tennessee, there's no guarantee that nationwide women's suffrage would have passed until decades later. Surely, American women would have somehow gotten the vote by the time France, Italy, and numerous other nations fully enfranchised their adult female citizens after World War II. But without Harry Burn's vote in Tennessee in August 1920, votes for women might have been delayed in the United States by a decade or two.

Compare the narrow success of the Nineteenth Amendment to the fate of its sister legislation, the Equal Rights Amendment, written by suffragist leader Alice Paul, and first introduced in 1923, only a couple years after women's suffrage became national law. In the early 1940s, the measure was so popular that both Republicans and Democrats added support for the ERA to their party platforms. The ERA passed Congress in 1972 and was sent to the states for ratification. Yet, after nearly a century, the amendment remains unratified—by its deadline in 1982, the measure had gained 35 of the 38 states needed, falling just three states short.[6]

Good climate legislation has had close calls in the past too. As Nathaniel Rich explains in his book *Losing Earth*, in 1979, the Carter Administration commissioned a report confirming that climate change was dangerous and recommending that the federal government act quickly to cut greenhouse gas emissions. Over the next decade, scientists worked with policymakers to come up with a response at the scale of the threat, but the politics were too difficult to solve, and by 1989, the opportunity had been lost.[7]

Naomi Klein laments that this same period, just after the election of Ronald Reagan in 1980, was the start of a conservative revolution calling for unfettered capitalism with reduced government regulation, a rollback of environmentalism that came along at exactly the wrong time for the climate crisis. Free-market ideology ascendant in the United States and to a lesser extent in other wealthy nations doomed to irrelevance shortly after the ink on each agreement was dry international climate accords like Kyoto in 1997, Copenhagen in 2009, and Paris in 2015. It may seem that the world's governments have already squandered our best chances to avoid climate chaos. But a longer view of history shows that our best chance may in fact be coming right now.

Cycles of Crisis and Massive Change

Today is another period when serious climate action is possible. After four decades of domination, free-market economics is now under serious attack everywhere. In the United States, people of all walks of life are convinced that the rich have grabbed too large a share of the nation's wealth; that the economic playing field is not level and needs to be fixed; and that a good life means more than individuals making money and buying more stuff.

Covid was a wake-up call across the world and around the United States. A brush with death and months of isolation have helped those of us still here to reexamine our priorities and look for ways to make the economy work better to meet real human needs for meaning and connection rather than fulfill desires for luxuries and entertainment.

In fact, today may be a much better moment for climate action than any time in the last 40 years, according to predictions by forecasters with strong past records of success.

George Friedman is one of the world's leading analysts of world politics. His firm, Geopolitical Futures, publishes reports read in government ministries from London to Beijing as well as inside the Pentagon and on Wall Street. Over the last two decades, Friedman has bucked conventional wisdom to correctly predict that the United States would not be eclipsed as a world power by the rise of the European Union, that the rapid growth rate of China's economy would eventually slow, and that a long civil war would break out in Syria.

Friedman doesn't believe that history goes just one way, towards something new, whether it's improved or not. Instead, he takes a cyclical view of history, finding that periods of peace and prosperity alternate with periods of war and depression on evenly spaced and relatively predictable cycles.

There's nothing mystical about it. History repeats itself not by some unhidden law of physics or the hand of God or the inexorable operations of fate straight out of a Greek tragedy. Cycles recur because generations of people are raised in a certain kind of environment that gives them different experiences, abilities, and desires than their parents or grandparents. It's reminiscent of the old saying: "Hard times create strong men. Strong men create good times. Good times create weak men. Weak men create hard times."

It's not a kind of astrology to divide American history into eighty-year cycles as Friedman and other political analysts do. It's just a way of using the past to try to predict the future. The same idea popularized by those who examined the relationships of older Boomer and X generations to younger people in the Millennial and Z generations. In the 1990s, two of the best-known writers on the relationships among generations, Neil Howe and William Strauss, figured out that societies cycle through four 20-year "turnings"—High, Awakening, Unraveling, and Crisis—around every eighty years. During the Crisis period or "Fourth Turning," a major war, long period of civil unrest, deep recession or depression, natural disaster, or some combination of catastrophic events presents a threat to a society that its people perceive as potentially fatal.[8]

Like Strauss and Howe thirty years before him, Friedman agrees that 2020 was the year the current crisis period would begin. In his book *The Storm Before the Calm: America's Discord, the Coming Crisis of the 2020s, and the Triumph Beyond*, Friedman predicts that the decade of the 2020s will bring dramatic upheaval and reshaping of American government, foreign policy, economics, and culture. Such a period will offer many challenges but also an opportunity for positive change not possible until now.

"The United States periodically reaches a point of crisis in which it appears to be at war with itself," Friedman writes, "yet after an extended period it reinvents itself, in a form both faithful to its founding and radically different from what it had been." His theory is worth a closer look as it convincingly shows that now might finally be the time when America is ready for serious action on climate.

Surveying 250 years of history back to the American Revolution, Friedman finds that certain historical patterns have repeated themselves every few generations. For example, there have been four political cycles since the United States declared independence from Great Britain, each one beginning with a crisis that resolved problems that had been brewing for years.

The first crisis was the Revolutionary War itself, a crisis that grew out of demands for self-rule that American colonists had been making for the previous two decades since the French and Indian War. To justify throwing off the king and noble overlords in Britain, the American revolutionaries challenged the prevalent idea that the natural order of society was a hierarchy based on birth. Instead, in 1776 the Declaration of Independence asserted that "all men are created equal."

America's founders didn't just change their political system to replace colonial rule with home rule. They also changed their social system from aristocracy to democracy, at least in theory. Unfortunately, equal rights were reserved for white men with property. All others, including women, Black people, indigenous people, and even white men without property were excluded. This sowed the seeds for a new period of conflict decades later.

But first came a long period of peace and prosperity, the calm after the storm of Friedman's title. But the resolution of the issues from the Revolution created a new set of problems. The system could solve some of those problems peacefully. For example, white men without property were given the vote and access to political power. Other issues would have to wait until a later period, especially votes for women, as we saw earlier.

The problem that this system could not solve was how to square democracy and equality on the one hand with slavery on the other. One political compromise after another between those who wanted to expand slavery to new territories in the West and those who increasingly came to abhor slavery released tension for a few years until the next controversy came along. Ultimately, American politics was entirely eaten up by the question of whether to expand slavery or not. Tensions escalated from heated debates in the press in the 1830s to a flood of enslaved people fleeing from South to North in the 1840s to fistfights on the floors of Congress in the 1850s to the first shot fired in the Civil War in 1861.

As it so happens, this cycle from peace through brewing conflict to civil violence took eighty years to complete.

Due to the length of generations and how long each one was born after the most recent crisis Friedman finds that it generally takes about eighty years for another crisis to erupt. So, eighty years after the Revolution came the crisis of the Civil War. Another eighty years later came America's third existential crisis, the Great Depression and World War II.

Finally, eighty years after that, Friedman predicts that the 2020s will bring another period of upheaval that will mark the end of one era and the beginning of a new age. While some of the biggest developments in American history such as economic growth or increase in population have moved forward and have gotten bigger over time without cycling back to an earlier level, other key factors in our history have indeed seesawed back and forth, giving credence to Friedman's approach of identifying repeating cycles in history.

In politics, periods of progressive reform have alternated with other periods of conservative retrenchment. Likewise, the economy, while growing overall, has done so through periods of expansion and contraction. Nothing is more common in the financial media than to look back at past cycles of boom and bust, bull and bear markets, and prosperous eras closed by recession, depression, or an old-fashioned stock-market panic.

Friedman's book came out in early 2020 and doesn't mention Covid. But it's easy to see how the outbreak of the pandemic could be the trigger for the decade of crisis that Friedman predicts. A period of crisis puts many things in flux that seemed solid previously. It's a frightening time of political and economic turmoil when both political and social structures begin to shift. "Many see it as a sign that the country is coming apart, but in truth it is simply evidence of a rapidly evolving country passing through an orderly change," Friedman writes reassuringly.[9]

Crisis brings intense suffering to some and instability to all as it creates a freedom to solve problems that society's elites were able to block before. It is hard to think of a problem that qualifies as deferred by elites more than climate change, except perhaps for racial discrimination and wealth inequality. Though Friedman writes little about climate as a problem and is skeptical of international agreements, he does say that race is a unifying factor across American history that returns again and again especially at periods of political or economic crisis and transition.

Fifty Years of Reaganomics Coming to an End

Along with a political cycle lasting eighty years, Friedman finds another cycle of fifty years that determines the economy. What will make the 2020s so turbulent is that this is the first time since the founding of the United States that both cycles will end at about the same time. The nature of the economic cycle is especially applicable to the problem of climate change as it explains why so little action has happened so far.

Indeed, if we believe Friedman, we can see that the "lost decade" of the 1980s bemoaned by both Nathaniel Rich and Naomi Klein as a tragic missed opportunity for climate action was precisely the time when the United States was least likely to act on climate.

No matter how much scientists may have urged serious action, the economy of the 1980s was set up to do the exact opposite of cutting back on fossil fuels or consumer products—it was set up to use more stuff and, as a byproduct, to create more pollution and waste. That's not because Ronald Reagan ushered in an unexpected conservative revolution that overthrew a liberal order that could've lasted for decades longer if Jimmy Carter had been reelected, at least according to Friedman. Instead, the New Deal-era economy had become unbalanced and lacked investment capital needed to add productive capacity to meet consumer demand. With his pro-business approach, Reagan was just the man for the times.

According to Friedman's fifty-year theory of economic cycles, due to its own growing imbalance between the rich and everybody else, the Reaganomics economy that began in 1980 is due to end around 2030. The new economic paradigm will reject the free-market approach of the current period with its tax cuts for the wealthy and service cuts for middle-class and low-income Americans.

Friedman predicts that the transition will be a time for redistributing wealth and increasing opportunity for those left out of today's economy. The politics behind the transition will involve an unlikely alliance crossing both race and party lines between those who feel most left out of an economy dominated by finance and high tech, "who face a grim future without a significant change in their circumstances," but are still young enough to do something about it: people in the early stages of their careers who find their prospects blighted. "In a sense, these will be the millennials who do not live in Manhattan or San Francisco and do not work in marketing or high tech. They are the millennials who don't fit the cliches of the dominant culture."

Adding up members of the white working class with most Black Americans will yield just under half of the national population, a voting bloc that cannot be ignored. "It is a powerful coalition bound by common interests and not sentiment. It will be a strange and uneasy alliance, of course, but one with precedence."[10]

The New Deal was one of those precedents, a time when white workers in Northern cities and white farmers in the South came together with Black Americans across the country to support FDR's ambitious job and social programs (though once the programs were passed and implemented, Black workers were often left out). This is promising for today's multiracial alliance for climate justice, suggesting that a period of economic transition will also be a time to address longstanding racial inequities.

The economic transition of the 2020s will not proceed any more smoothly than the transitions between the previous political cycles in American history, says Friedman. Those political cycles all started with a crisis: the American Revolution, the Civil War, and World War II.

Past economic cycles may or may not have overlapped with a shooting war, but they did start with an ideological battle between partisan contenders where one side emerged with a decisive victory over the other. Thus, at the beginning of the Reaganomics cycle, the idea of the previous New Deal cycle that government should manage the economy to provide jobs and prosperity for the greatest number, was firmly rejected in favor of a new paradigm of "small government" and tax cuts for the rich that would encourage private investment.

That era of encouraging private investment is now ending, according to Friedman, because there's too much investment capital in the economy with too few opportunities in which to invest, yielding the historically low interest rates of the last few years. And the 2020s will be a period of economic as well as political contention.

On the economic side, Reaganomics will make its last stand, and during the decade, a president will be elected who will try one last round of free-market measures including tax cuts for the rich, just as Herbert Hoover did to try to fix the Great Depression just before FDR came in. This will backfire, Friedman predicts, and the next president will take a new approach focused on solving the problems of the mass of middle-class Americans at the expense of the rich, rejecting Reaganomics once and for all.

Will the new approach hearken back to the job and service programs of the 1930s New Deal?

Friedman leaves that opportunity open. But he does suggest that what must be different in the future from the past is that government will need to become less complex, less about following bureaucratic rules and more about giving ordinary citizens access to redress. The original Social Security Act passed under FDR was less than 100 pages in length. By contrast, the Affordable Care Act ballooned to more than 20,000 pages of regulations put out by the Obama White House by mid-2013, a pile of papers that would reach nearly six feet high if you made the mistake of trying to print it.

This is just one example of how our government has become too complicated. Government officials don't understand everything about their roles and are empowered only to execute a narrow part of a much larger system, which leads to waste and ineffectiveness. Ordinary citizens can't comprehend how government works and thus can't take full advantage of its benefits or fight against its threats. Only the wealthy, who can afford attorneys and lobbyists to navigate the complexities of regulations, get full benefit from a system so complex. Some would argue that government needed to become more complex in response to an economy based on increasingly advanced technology and dominated by big corporations.

Whatever the cause, for the average citizen, it's become nearly impossible to understand how to address an ongoing grievance or make significant change, whether government or corporate. Only the rich can hire the high-priced professionals it takes to know how the system works so that they can work the system.

The New Economy Must Be Both Clean and Fair

The Green New Deal or the Just Transition from 350.org will need to avoid the mistakes of Obamacare and pursue a simpler, more human-scale intervention in the American economy. Though the goal is transformational change on a massive scale, the method must use a decentralized and more distributed model of eco-friendly development not dependent on economic growth, the kind of economy suggested by many experts in the field of ecological economics.

Economist Kate Raworth, for example, says that over the last few decades, economic growth has failed to live up to its promise to float all boats and to distribute prosperity widely. "No economist from last century saw this picture, so why would we imagine that their theories would be up for taking on its challenges?" Raworth asks, calling for an economics suited to the needs of the current crisis.

> We need ideas of our own, because we are the first generation to see this and probably the last with a real chance of turning this story around. You see, 20th century economics assured us that if growth creates inequality, don't try to redistribute, because more growth will even things up again. If growth creates pollution,

don't try to regulate, because more growth will clean things up again. Except, it turns out, it doesn't, and it won't. [11]

Growth has failed both to offer a decent living to everyone but the top one percent while industrial production has bumped up against ecological limits—using up supplies of finite resources like metals while pumping out pollution beyond the ability of the air, water, and earth to absorb it. Thus, the climate crisis. But also, a crisis in equity.

We can't allow an economy that has failed to provide a good living to most people alive to simultaneously destroy the planet. "Humanity's twenty-first century challenge is clear: to meet the needs of all people within the means of this extraordinary, unique, living planet so that we and the rest of nature can thrive," writes Raworth.

The future economy must reach a state of dynamic equilibrium between the ceiling of the limits to the earth's resources and the floor of the minimum prosperity needed to provide a comfortable living for all people everywhere. Though she comes from a different perspective than Friedman, Raworth also believes that the next few years will offer a unique opportunity for economic reform.

It seems that even members of the top one percent agree. Wealth disparities have grown so obvious and so obscene that even plutocrats themselves have started to worry about political instability. Some wealthy people, including Bill Gates and Warren Buffet, are trying to defuse the wealth disparity bomb by calling for reforms to distribute wealth more equitably such as raising taxes on the richest people. Other rich people see no hope of averting a revolt of the masses. They've put their pilots on alert and fully gassed up their Gulfstream 550s ready to take off on half an hour's notice for their bug-out retreats in New Zealand.

If Raworth and Friedman are correct, the decade ahead will likely be one of turmoil, unrest, and ultimately, transition. This means that serious action will have a much better chance now than it has in the last fifty years. Ecologically, it would have been better to cut greenhouse pollution back in the 1980s and 1990s when changes could have been made more cheaply and in time to prevent much human suffering and damage to the physical environment. But perhaps no amount of science would have made serious climate action possible given the political and economic orthodoxy of the time.

Today is already different, and the next few years are likely to become even more receptive to a much different kind of politics and economy. That's what happened in the 1930s. The laissez-faire orthodoxies of the Gilded Age and the Roaring Twenties failed to prevent economic collapse or put Americans back to work. Old ideas about the economy were discredited and rejected in favor of muscular action by government on a scale so transformative as to qualify as a revolution.

A similar period of revolutionary change came about eighty years earlier, at the time of the previous existential crisis that threatened to rip America apart, the Civil War. After decades of contention, the issue that had been endlessly deferred and compromised about, slavery, finally took its rightful place at the center of the nation's politics.

The Abolition of Slavery Was Not Inevitable

Now, let's return to the commonly held theory that slavery would have just peacefully faded away as the economy naturally shifted from farms to factories, from rural areas to cities and suburbs as part of the momentum of the Industrial Revolution.

Modern people may think that slavery's end in the United States was inevitable at some point in the nineteenth century, even without the Civil War. After all, the British Empire had freed its slaves in the 1830s and France ended slavery in its colonies in the 1840s. With the rise of industry and the decline of agriculture in the mid-nineteenth century, so the story goes, American slavery, which was never very profitable anyway, was ready to wither away and collapse on its own.

This story may be false. Recent histories paint a very different picture of the economics of slavery and suggest that predications that slavery would have eventually withered away and then died by natural causes were so much wishful thinking.[12] As it turns out, slave-powered commodity agriculture was profitable up to the very end and was ready to expand right up to the day in April 1861 when secessionist troops fired on Fort Sumter and ignited the Civil War.

If the Confederacy had been a separate nation right before the war, it would have ranked as the fourth largest economy in the world. The fast-growing Deep South states had a higher proportion of rich people than New York City. Just as oil is today, so was cotton the most widely traded commodity on earth at the time, when textile mills led the Industrial Revolution in both the U.S. and Britain. Seventy-five percent of cotton traded worldwide was grown by enslaved labor in the Southern states.

According to historian David Blight, "by 1860, there were more millionaires (slaveholders all) living in the lower Mississippi Valley than anywhere else in the United States. In the same year, the nearly four million American slaves were valued at $3.5 billion [about $10 trillion in today's money], making them the largest single financial asset in the entire U.S. economy, worth more than all manufacturing and railroads combined."[13]

Most of the wealth held by Southerners was in enslaved people, who could fetch up to $2,000 each on the auction block, which would equal about $62,000 in 2020 dollars.[14] But everything else that Southern planters and capitalists owned was enhanced by slavery, including their plantation land, farming equipment, and port facilities.

Before the Civil War, the issue driving slaveowners was not simply whether they could keep their slaves. It was whether slavery could expand to new territories. In 1850, a decade before the war, Southerners successfully pushed the federal government to enact a harsh Fugitive Slave Act. This divisive law didn't just allow slave catchers to pursue escaped Black people into free states, but it also required any citizen in the North to help capture self-emancipated people and return them to Southern enslavers. The Fugitive Slave Act essentially legalized slavery nationwide, even in Northern states that had abolished it years earlier. The new law was a hollow victory that gave Southern planters little benefit but spread anger at slavery across the North.

In the years leading up to the Civil War, political leaders from the South led calls to annex Texas and then fight a war with Mexico to acquire the whole southwest to open new territory to slavery. But even that wasn't enough to satisfy King Cotton. Southern dreamers had plans to expand slavery even further, south into Central America and the Caribbean.

If the South had won the Civil War, slavery would have continued in all the states of the Confederacy for decades if not longer, while possibly also spurring wars of conquest to give slavery even more room to grow. Imagine if the Spanish-American War of 1898, where the United States wound up occupying Cuba and annexing Puerto Rico, had instead been the Spanish-Confederate War, where Southern planters wound up taking those islands and other Spanish colonies to create a growing empire of slavery? It's easy to imagine that slavery would have lived on by moving into other tropical areas of the hemisphere where even today farming relies more on human labor and less on farm equipment than in the U.S. and other wealthy nations.

It may sound like a crazy idea to think that without the Civil War, slavery might have persisted in the Americas for decades longer and even into the twentieth century. Sometimes we need to turn from history to art to imagine an alternate outcome to past events. The 2016 novel *Underground Airlines* by Ben Winters makes the implausible sound almost natural.[15]

The story imagines an alternate American history where Lincoln was assassinated not in 1865 at the end of the Civil War but in 1861, before the start of his first term. In the novel, the whole nation is so horrified by the president's death that Southern states which had already seceded come back into the Union on condition that an amendment is passed to forever protect slavery in the South from abolition by the federal government, thus avoiding the whole Civil War.

Fast forward a century and a half to the present day in the plot of *Underground Airlines*, and all but a handful of Southern states have abolished slavery on their own. But forced labor remains legal in the "Hard Four" states of a unified Carolina, Alabama, Mississippi, and Louisiana. In a high-tech industrial economy, those states run an updated version of slave plantation agriculture and manufacturing. In open-air work camps like modern state penitentiaries surrounded by razor wire and guard towers, Black Persons Bound to Labor, known as "peebs," dressed in prison-style orange jump suits and monitored by microchip implants, work farms and factories, producing cheap clothing for multinational corporations such as Garments of the Greater South, Inc.

Winters's vision is so frightening because its use of details from globalized manufacturing shows that slavery may be more compatible with high-tech capitalism than anybody thought. Fortunately, America never had to test this theory. But if we hadn't abolished slavery after the Civil War, it might have set an example for the other places in the Western Hemisphere that still allowed slavery in the late nineteenth century, Cuba and Brazil. Brazil was the last country in the Americas to abolish the institution, waiting until 1888 to free its enslaved people. If the American South still allowed slavery into the same period, perhaps the Brazilians would have found good reason to postpone their emancipation too.

Abolition was a triumph, but an incomplete one. The heartbreaking fact remains that in the present day more than 40 million people are enslaved worldwide. That's more enslaved people than there were on the planet in 1860.[16]

Contemporary slavery doesn't look like slavery did in the nineteenth century. Since slavery has been outlawed by every nation on earth— Mauritania was the last to abolish slavery in 1981, though enforcement has been weak, and the country still has many people held illegally in bondage—today's slave economy must operate underground, trafficking women and children as forced sex workers, children as soldiers, or workers in diamond mines or on plantations that produce coffee or cocoa beans. To end illegal slavery, activists have mounted campaigns to certify "slave-free chocolate" and eliminate "conflict diamonds."

Indeed, the work of abolition remains undone, both in terms of actual illegal slavery around the world and, in the United States, of violence against and unfair treatment of Black Americans. The persistence of these challenges does not tarnish the luster of what Black and white abolitionists accomplished in 1865. After nearly a century of agitating for freedom against the growing financial and political might of the Slave Power over the federal government, the crisis of the 1860s offered a time that was finally right for success.

History also shows that today may offer another period of crisis that first dislodges and then casts out to sea the ideology of a dying era. Today's crisis may offer the ripest time ever for victory by the greater climate justice movement, an alliance interested not only in cutting greenhouse gas emissions to safe levels but also in cutting out racial discrimination and the biggest threats to public health, while greatly reducing the gap between the top one percent and all other Americans.

3

Dare Not Make War on King Cotton or Big Oil

What would happen if no cotton was furnished for three years? I will
not stop to depict what everyone can imagine, but this is certain:
England would topple headlong and carry the whole civilized world
with her, save the South. No, you dare not make war on cotton. No
power on earth dares to make war upon it. Cotton is king.

—James Henry Hammond, senator from South Carolina, speech on
the Senate floor, 1858

American business prides itself on leading the world in
innovation. Yet, all too often, the biggest businesses threaten
that if society asks them to do things differently, catastrophe will
ensue not just for their business but for the whole economy. This can't-
do spirit has become all too common in American industry over the last
half century.

In the 1960s, for example, General Motors and other American
automakers claimed that making them install seat belts would force
them to raise the price of cars beyond what consumers could afford.
That in turn would lead to lower sales, forcing automakers to close
plants and lay off workers. Scary…but it never happened. As it turned
out, automakers did just fine selling cars with seat belts included, as they
did with other innovations demanded by the public which car
companies also originally opposed, from air bags to better gas mileage.
But that hasn't stopped automakers and other big industries from

continuing to complain that having to make their products safer and less polluting would spell disaster for their companies, their workers, and the whole economy.

This excuse didn't start with automakers in the 1960s. Resistance from big business about having to reduce the harmful side effects of their products to workers and consumers goes back more than a century, to debates over everything from garment sweatshops at home and abroad; pollution from DDT in water supplies; monopoly railroads charging farmers high rates for freight; and of course, slavery.

In the decades before the Civil War, when campaigns by abolitionists started to reach a wider audience, Southern cotton planters complained that all the criticism was harming their business. So, big planters and the politicians they controlled responded with a public relations campaign whose main message was that any threat to the expansion of slave agriculture would be a threat to the whole economy. And not just the economy of the South, but the economy of the whole United States and even of the larger Atlantic world.

When South Carolina planter James Henry Hammond warned his colleagues in the U.S. Senate that they shouldn't make "war" on King Cotton, he wasn't referring to a shooting war, yet, though three years later his fellow South Carolinians would fire on Fort Sumter and start the Civil War. But in 1858, Hammond's state was still in the Union, and at the time he meant that if the federal government presumed to restrict the spread of slavery to new territories out West, then it would depress slave-state cotton exports and spell disaster for the closely entwined economies of America and Great Britain.

In the mid-nineteenth century, the Industrial Revolution was gaining steam and Britain was the leader in building factories and railroads, making it the world's richest economy and the world's leading military power.

Cotton was America's biggest export and British fabric mills were the biggest buyers of Southern cotton, followed by makers of cloth in New England states. If Southern planters were forced to make do with only the territory they already had without being able to keep opening new places to slave agriculture, so the argument went, cotton production would fall behind rising demand. Lack of supply would cause cotton mills on both sides of the Atlantic to curtail production or even shut down entirely, putting workers from Manchester to Massachusetts out of a job and spiking prices of fabric and textile products on both sides of the Atlantic. The effects of crippling one of the biggest industries in both Europe and America would spread from unemployed workers who couldn't pay their rent to middle-class landlords who couldn't pay their mortgages to wealthy bankers who couldn't meet their own obligations to move money around the economy.

After they seceded to form the Confederacy, Southern leaders got the chance to put this theory to the test. They tried to use cotton for diplomacy. Confederates wanted to pull Britain into the American Civil War on their side.

Since half of Britain's foreign trade and one in four British jobs were dependent on cotton, 80 percent of which came from the American South, Confederate leaders figured they'd have Britain over a barrel if they put pressure on the British economy by restricting the supply of cotton. So, near the beginning of the war in the summer of 1861, the rebels predicted that they'd have everything to gain and nothing to lose when they declared an embargo on cotton exports to Britain and other European powers.

The Confederate government ordered 2.5 million bales of cotton burned to create an artificial shortage of supply. Initially, the impact of this bold move was limited. Since British importers had been stockpiling supplies for a couple years in anticipation of war, clothing mills could keep running for a few months even with new supplies of raw cotton cut off. But by the end of 1862 the Northwest region of England, the epicenter of its textile industry, suffered a "cotton famine" that put 500,000 out of work, leading to food riots. Southern leaders insisted that this suffering would cause Britain to recognize the Confederacy and then join with other European powers to pressure Lincoln to make peace on the South's terms.

What could possibly go wrong?

Britain valued peace with the United States and did not want to risk war by openly supporting the Confederacy, though British shipbuilders did quietly outfit Confederate warships and British merchants did trade with Confederate blockade runners. But the government's official policy remained non-intervention and British mills found a way to remain open through the wartime loss of American cotton supply, seeking alternate sources in India, Egypt, and Brazil. This thwarted the Confederate plan to use cotton for strong-arm diplomacy and showed that you could in fact make war on King Cotton and win. Britain stayed out of the conflict and the Union prevailed over the Confederacy.

Cotton's power was broken and would never be restored. Even after the destruction and expense of the Civil War, the United States economy boomed after abolishing slavery. When peace came and slavery was abolished, cotton was still grown in the South, but this time under free labor, though not as free as it should have been, as we'll see later.

Southern planters started shipping cotton overseas again, but the crop was no longer America's leading export, merely a sideline in a rapidly industrializing economy. The post-war Gilded Age, a period of industrial growth when America first became a world power, became famous for tycoons like Andrew Carnegie in steel and John D. Rockefeller in oil and prolific inventors like Thomas Edison. Despite heated battles between labor and management, the era's economy was so hot that it attracted wave after wave of immigrants, the much celebrated "huddled masses yearning to breathe free" who became the ancestors of so many Americans today.

The Slave Power was wrong that abolishing slavery would destroy the U.S. economy in the nineteenth century, and the Oil Power is equally wrong to claim that abolishing fossil fuels will harm the U.S. economy in the twenty-first century. Throughout American history, whenever the public has tried to clean up big dirty businesses that oppress people, whether Big Slavery or Big Oil, their wealthy owners have always tried to claim that their private interest is the same as the public interest, that the public dare not "make war" on their industry. But invariably, the exact opposite is the truth.

When the government reins in industries that are either literally dirty or morally dirty or both, not only do workers and communities stand to benefit, but the economy often booms as well. The booming solar industry, the fastest growing source of new jobs in today's economy, is just one example of how a clean energy revolution will increase prosperity. Plans for the Green New Deal draw on dozens of studies from economists projecting new jobs and opportunities for entrepreneurs.

Back before the Civil War, slavery was a violent way for a few rich white planters to extract wealth from the forced labor of Black Americans, trampling on their rights to life, liberty, and happiness. But slaveowners tried to spin their oppressive business model as the best arrangement for everybody, whether the enslaved themselves or buyers of products made affordable by cheap labor. Enslaved people and free abolitionists alike always knew that was a huge lie.

Today, when we have clean alternatives to generate power and increasingly affordable batteries to store that power for future use, oil has become just another way for a few rich executives and investors to extract wealth from the earth and from our pockets, trampling on the rights of present and future generations to a livable world. Yet, after decades of rapid growth by clean energy, the dirty energy industry still claims that abolishing their deadly product would lead to dire consequences. The truth is that any dire consequences from abolishing oil are most likely to hit oil company executives and their investors harder than anyone else.

Claiming that harmful ways of getting rich are in the public interest is just one of the myths that immoral industries always spread to try to protect their profits.

Doubt Is Their Biggest Product

If you add up all the reasons that fossil fuel companies have given over the last 30 years for why the economy dare not try to get off oil, gas, and coal, the excuses start to cancel each other out.

If climate change is a hoax, then why does it matter if power plants switch from dirty coal to "clean" coal? If pumping out greenhouse gases is not the fault of the companies that produce and sell oil but instead is really the fault of billions of consumers driving their cars and shopping at Amazon, then why does it matter if oil companies are now reducing their carbon footprint? If it will be too expensive to build enough solar panels and wind turbines to replace dirty energy over the next three decades, then how is it that the public can still afford to give billions of dollars in subsidies every year to oil, gas, and coal companies?

One reason why dirty energy propaganda contradicts itself is that fossil fuel companies' messages have changed over time in response to conditions in both the climate and in politics. The climate has gotten worse, scientists' warnings have gotten more urgent, and voters have gotten more worried. To accommodate this new political environment, the things that oil and gas companies say about climate sound nicer now than the things they used to say. Those companies used to just deny the science and attack the scientists. Now, the companies admit that there's a problem and say they want to be part of the solution. To prove it, many oil companies have promised to make their operations carbon neutral or achieve "net-zero" emissions by 2050 or sooner.

But we shouldn't be fooled. The dirty energy lobby is still waging war on the climate. Big Oil has just updated their approach. Instead of wielding a bare iron fist out in public like they used to, they now make sure to slip on their velvet gloves before leaving the house. But it's a fist all the same.

In his book *The New Climate War*, scientist Michael Mann explains that the old climate war was straight-up science denial by fossil fuel companies to delay any government action that would restrict coal, oil, and natural gas. This political strategy worked so well, according to Mann, that "we are now witnessing the devastating effects of unchecked climate change." Since coastal inundation and flooding along with record-breaking heatwaves, long droughts, and out-of-control wildfires are now staples of TV news, newspaper headlines, and social media posts, it's no longer credible for the forces of denial and delay to claim that climate change isn't happening or that it's not dangerous.

Even for fossil fuel companies, right-wing plutocrats, and oil-funded governments, Mann writes that "outright denial of the physical evidence of climate change simply isn't credible anymore. So, they have shifted to a softer form of denialism while keeping the oil flowing and the fossil fuels burning, engaging in a multipronged offensive based on deception, distraction, and delay. This is the *new* climate war, and the planet is losing."

Everybody's got a right to change their mind and what they say in response to new information. The problem with dirty energy companies is that they may be saying that they accept climate science and want to be part of the solution, but their actions show that the words are still just spin. "When it comes to the war on the science— that is, the *old* climate war—the forces of denial have all but conceded defeat. But the new climate war—the war on *action*—is still actively being waged," Mann explains.[1]

In the end, it doesn't matter that much what Big Oil says, because their history of deception has given them very low credibility. And they seem to be willing to accept that. If you're looking for a coherent story from 30 years of PR from the fossil fuel lobby, then you're missing the point. Big Oil does not need to convince the public that what they say is true. They just need to spread enough fear, uncertainty, and doubt about climate science and climate solutions to discourage society as long as possible from taking any action that would hurt their profits.

"Doubt is our product," a PR executive for a cigarette manufacturer once observed, "since it is the best means of competing with the 'body of fact' that exists in the minds of the general public. It is also the means of establishing a controversy."[2]

Starting in the 1950s, when health researchers began issuing warnings that smoking caused cancer, tobacco companies ran a successful PR campaign to cast doubt on this research, managing to postpone government restrictions on cigarette sales for 30 years. A couple decades later, when scientists started issuing warnings about climate change, fossil fuel companies took a page out of the tobacco PR handbook. They launched a campaign to manipulate public opinion to buy themselves more time to pollute.

As early as 1959, the physicist Edward Teller told a meeting of the American Petroleum Institute that a ten percent increase in CO_2 from burning fossil fuels would be enough to melt the Arctic icecap and submerge New York City. "All the coastal cities would be covered, and since a considerable percentage of the human race lives in coastal regions, I think that this chemical contamination is more serious than most people tend to believe."

Less than a year later, the API received a report it had commissioned from researchers at Stanford University with a direct warning on the dangers of carbon dioxide: "Significant temperature changes are almost certain to occur by the year 2000, and these could bring about climatic changes…there seems to be no doubt that the potential damage to our environment could be severe… pollutants which we generally ignore because they have little local effect, CO_2 and submicron particles, may be the cause of serious world-wide environmental changes."[3]

Exxon claims that, though people in their company had heard of this research, the science was still unsettled and not yet conclusive enough to act. "News reports that claim we reached definitive conclusions about climate change decades before the world's experts are simply not accurate," said a company statement in response to coverage in the *Guardian*.[4]

By now, fossil fuel companies surely understand that they will have to shut down someday. If nothing else, their reserves of oil, gas, and coal will peak and become too expensive to produce. But before that, citizens may finally get organized enough to use their power to demand that government force dirty energy companies to leave fossil fuels in the ground. That's the outcome that Exxon, Koch Industries, and pipeline developers want to put off as long as they can. In their battle against public accountability, delay counts as victory, and success for fossil fuel interests looks like another fifty years of selling dirty energy as cheaply as they can.

So, all the excuses they offer and the myths they spread aren't meant to add up to a convincing argument. They're just meant to confuse the public enough so that we won't take any meaningful action anytime soon. Recent pledges to achieve "net-zero emissions" in their operations are just a delaying tactic. "Not only is the oil industry unlikely to be a leader on carbon reductions, but the sudden flurry of net-zero pledges is instead a cynical effort to bolster corporate images in a calculated attempt to buy time to extract more oil and gas," as *DeSmog* explains.[5]

Oil companies used to have a much unfriendlier attitude to climate science. Frankly, their former approach to climate was much more straightforward. Before we take apart their current more complex way of spreading misinformation by claiming that they can be part of the solution on climate, it will help to survey the good old ways that fossil fuel companies used to lie about climate science in the simpler days of yore.

Michael Mann refers to straight-up denial as the "old climate war," as opposed to the "new" climate war, a stealth attack on the climate where companies pretend to care about science and try to play nice while still planning to produce as much fossil fuel as they can, clean air be damned. But even if the biggest companies have moved on to new tactics, many of their allies in conservative media and propaganda shops continue to stick with the old lies, showing that climate denial hasn't gone entirely out of style.

Their excuses bear a striking resemblance to the excuses used by slaveholders at various points in the eighteenth and nineteenth centuries.

Numerous writers on climate change have catalogued and rebutted the myths spread by fossil fuel companies and their allies about why it's not practical, possible, or even necessary to transition the whole economy to 100 percent clean energy in the foreseeable future. Here, we'll see a summary of each claim in a new context.

In fact, it's an old context, but one that's new to the conversation around climate change: we'll connect each excuse made by fossil fuel companies in recent years to the same kind of excuse from big cotton planters before the Civil War for why they couldn't free their slaves voluntarily and why society shouldn't force them to do it. Comparing six major lines of argument by Big Oil with similar excuses by Big

Slavery makes clear the moral and logical bankruptcy of each excuse. The comparison also suggests that dirty industries like cigarettes and fossil fuels didn't have to reinvent the wheel when it came to waging war on the facts. Modern PR and lobbying shops could just copy and paste from the defenses of their forebears, going all the way back to King Cotton:

1. The science is wrong

2. Big change will be impractical

3. Poor people will be hurt the most

4. We can clean up our act

5. We're not to blame—you are

6. Your solution is too expensive, and it leads to tyranny

1. The Science is Wrong

First, since at least the 1960s, fossil fuel companies' own scientists have been telling them that burning hydrocarbons would lead to the greenhouse effect and potentially dangerous levels of climate heating.

As in so many problems caused by Big Oil, ExxonMobil offers an example. Over the following years, further warnings came from researchers both outside and inside the company. By the late seventies, the science had become more settled.

In July 1977, the company's senior scientist James Black informed management that scientists had concluded that humans were likely altering the environment by releasing CO_2 from burning fossil fuels. The next year, Black warned Exxon that doubling carbon in the atmosphere would raise the temperature by an average of two or three degrees Celsius, which is consistent with today's science. "Present thinking holds that man has a time window of five to ten years before the need for hard decisions regarding changes in energy strategies might become critical," Black concluded, saying that Exxon needed to act.[6]

During the 1980s, the company did invest in more research, even outfitting a tanker ship at a cost of $1 million to study how much carbon dioxide would be absorbed by the oceans. But Exxon didn't disclose the information they gathered to the public or to government regulators. They kept it to themselves.

Finally, in the late 1980s, it appeared that Congress might pass legislation to regulate carbon dioxide. At that point, Exxon and other oil companies decided that they needed to speak up. But instead of playing a constructive role in the conversation about climate, they decided to work against the public interest by starting PR campaigns to cast doubt on climate science, a battle plan they've been following ever since. Over the last 30 years, Exxon and other fossil fuel companies have spent billions of dollars trying to convince voters and political leaders that climate science wasn't settled, and that government should wait and see before taking any action to limit fossil fuels.

In the early 2000s, some oil, gas, and coal companies started to admit that burning their product was dangerous to the climate. That's not because they suddenly grew a conscience. It's just because they could no longer plausibly deny science in public. "What we are now seeing is a broad rhetorical shift by the fossil fuel regime towards more subtle and subversive tactics for delaying climate policies: a combination of misleading public relations and backroom lobbying that greenwash the industry's image while undermining meaningful action," said Harvard researcher Geoffrey Supran. "But it's denial by any other name. Because although the rhetoric and tactics have evolved, the goal—and the result so far—remains the same: inaction."[7]

So, while in public oil companies now say that they accept climate science and will try to be part of the solution, behind the scenes, these companies continue to support shadowy front groups like the Heartland Institute along with oil-patch politicians like Oklahoma's Republican Senator James Inhofe to spread lies about climate science. Inhofe once called climate change "the greatest hoax ever perpetrated on the American people," and later threw a snowball in the Senate chamber which was supposed to make some point about how a cold winter disproves climate science.

Oil company CEOs are too buttoned up to pull dumb stunts like Inhofe's. And these days, Big Oil is trying to convince the public that it has stopped denying climate science. Yet, despite their new messaging about climate, the energy and natural resource sector, mostly oil and gas companies, still gave Inhofe $1.08 million in the 2020 election cycle.[8]

As Michael Mann has explained, under the revised plan of attack developed by Big Oil and their allies for their new war on the climate, outright science denial has been replaced by more subtle tactics. But the oldest dinosaurs like Inhofe and the most rabid junkyard dogs in the oil patch like the Heartland Institute obviously didn't get the memo.

Or maybe they're not trying to reach the public but are only speaking to a small group of right-wing extremists. Heartland, for example, continues to hold a national conference every year for climate science deniers. Their 13th annual conference was held in July 2019 at the Trump International Hotel in Washington, DC, featuring "the courageous men and women who spoke the truth about climate change during the height of the global warming scare," many of whom went on to advise the administration that "the human impact on climate is likely very small and beneficial, not harmful."[9]

And some other attack dogs continue to snarl, but just a little more gently. "Climate change is real but it's not the end of the world. It is not even our most serious environmental problem," writes nuclear power advocate Michael Shellenberger, who was once named "Hero of the Environment" by Time magazine. His 2020 book *Apocalypse Never: Why Environmental Alarmism Hurts Us All* has become a best seller among free-market types who enjoy having a champion with green movement credentials to downplay the climate threat without having to explicitly deny climate science.

It doesn't seem to bother free-market PR agents that Shellenberger, who boasts a master's degree in anthropology but has no special qualifications in climate science, is contradicted by 97 percent of climatologists who have said for years and continue to warn that climate change is the most dangerous threat to civilization today.[10]

Big Oil used to claim that climate change was a hoax, while some of their nastier allies still cast doubt on climate science. Likewise, King Cotton used to claim that it was a hoax to believe that Black and white people were equal, and that enslaved Americans deserved to the same rights as free Americans.

In the era of the Revolutionary War, Americans both North and South recognized that slavery did not belong in a new nation founded, as Jefferson put it in the Declaration of Independence, on the self-evident truth that all men are created equal. Some slaveholders of the late eighteenth century believed that it was only a matter of time before slavery would end, and in their writings they claimed that they welcomed that ending.

Take the example of three Founding Fathers from Virginia, all slaveowners. Patrick Henry referred to slavery in a letter as an "abominable practice...a species of violence and tyranny" that was "repugnant to humanity." The first motion that Richard Henry Lee made when he entered Virginia's House of Burgesses in 1759 was "to lay so heavy a duty on the importation of slaves as effectually to put an end to that iniquitous and disgraceful traffic within the colony of Virginia." Thomas Jefferson, who also introduced a bill in Virginia to end the transatlantic slave trade, confessed in 1779 that "the whole commerce between master and slave is a perpetual exercise of the most boisterous passions, the most unremitting despotism on the one part, and degrading submissions on the other."

Antislavery slaveowners might just sound like hypocrites to the modern reader but they represented the dominant politics of their day.

But by the early nineteenth century, after the invention of the cotton gin made it much more profitable to grow cotton, slavery boomed in areas hospitable to the crop. As Chris Hayes points out, "between 1805 and 1860, the price per slave grew from about $300 to $750, and the total number of slaves increased from one million to four million—which meant that the total value of slaves grew a whopping 900 percent in the half-century before the war." In the face of big money getting bigger, Southern planters started to change their tune about slavery.

Even as American political culture was moving towards more equality in the 1820s and 1830s—for example, the right to vote was extended from white men with property to all white men regardless of wealth—Southern planter politics was moving in the opposite direction, towards racial hierarchy. As they adapted their political philosophy to support their way of making money, some cotton planters were even ready to cancel the Declaration of Independence itself. "I repudiate, as ridiculously absurd, that much-lauded but nowhere accredited dogma of Mr. Jefferson, that 'all men are born equal'," said South Carolina Senator James Henry Hammond.[11]

"Slavery is a positive good," argued Hammond's sometime ally and political leader from South Carolina John C. Calhoun, known as the "cast-iron man" for his aggressive defense of the Southern elite against abolitionists. Appealing to the racial pseudo-science of the day, Calhoun claimed that "Never before has the Black race of Central Africa, from the dawn of history to the present day, attained a condition so civilized and so improved, not only physically, but morally and intellectually... It came to us in a low, degraded, and savage condition, and in the course of a few generations it has grown up under the fostering care of our institutions."[12]

Like Calhoun, surgeon Josiah Nott appealed to one of the popular fake sciences of the day. In his case, it was the pseudo-science of phrenology, which drew conclusions about human mental function from the shape of the head, an approach that was criticized even in its own time for lack of rigor in evidence gathering. That didn't stop Nott from applying phrenology's bogus claims to justify racially based slavery. "There is a marked difference between the heads of the Caucasian and the Negro, and there is a corresponding difference no less marked in their intellectual and moral qualities," wrote Nott, concluding that the natural inferiority of Black people marked them biologically to take a second place in society to white people.[13]

To protect their wealth in human chattel and their profits from the cheapest labor possible, the slaveowners of the South used fake science to give birth to the ideology of white supremacy, a dangerous lie that would continue to spread its venom through American society long after their plantations were gone with the wind. This foreshadowed how coal, oil, and gas companies would later spread their own fake science to promote the free-market ideology necessary to shield their oppressive and polluting industry from reform.

2. It's Impractical to Change

Until recently, dirty energy lobbyists warned that it would be impractical or too expensive to switch to clean energy. Now, soundly trounced by reality over the last few years when solar and wind power have become the cheapest energy in most places—even when having to compete against highly subsidized oil, gas, and coal—Big Oil has changed its tune.

In the wake of the collapse of the world market for American coal and in the face of a national consensus that coal is too dirty for the twenty-first century, coal companies have nearly given up trying to talk to Americans. They're taking their sooty act overseas to countries like India where enough poor villagers are clamoring for electricity as quickly and cheaply as possible that they don't care how dirty it is—yet.

By contrast, Big Oil sees a chance to keep selling domestically as long as they can convince Americans that, even as we transition to clean energy, we can't possibly do without them at least until the middle of the century. In the next few years, petroleum companies claim, the economy will need natural gas to serve as a "bridge" fuel from old energy to new energy. As electric utilities retire their dirtiest power plants, usually those that burn coal, Big Oil claims that those utilities won't be able to jump straight into clean energy. Solar and wind supposedly aren't ready to provide the scale of power needed. Intermittent power sources (solar panels only work in the daytime and wind turbines only turn when the wind blows) can't provide always-on power needed to keep the electric grid up and running 24/7. Batteries would be needed to store power for the times that solar and wind aren't producing energy.

Until clean energy and batteries have a chance to mature, this argument goes, for the immediate future utilities will need to run power plants using natural gas. And fortunately, gas burns cleaner than coal, so oil and gas companies claim.

Switching from coal to gas is one reason America has been able to clean up its power sector, according to the industry. "U.S. carbon emissions have fallen to the lowest levels in a generation, thanks in large part to power companies transitioning from coal to natural gas—while CO_2 emissions from other countries have risen 50 percent since 1990," says the Energy for Progress website run by the American Petroleum Institute.[14]

Along with natural gas, oil companies predict that the economy will still need plenty of plastic, which is usually made from petroleum.

Though gas does emit less carbon than coal when it's burned, researchers have proven wrong the claim that natural gas is any better than coal for the climate. Once you consider methane leakage from wells and pipelines, natural gas may be even more dangerous for the climate than coal because methane is 30 times more powerful a greenhouse gas than carbon dioxide. As to the industry's claim that we'll also need oil in the future to make plastic, that's arguable as well.

Many products made today using plastic parts or constructed entirely from plastic could be made with alternative materials. Manufacturers can substitute metal, glass, cloth, and even advanced silicone for plastic. Or they can make bioplastic from crop products like vegetable fat, sawdust, and recycled food waste. This might raise costs initially, but in the absence of cheap oil, manufacturers would find a way to produce plastic alternatives more cheaply just as they have done in the past for other materials.

Plastic need not be used for disposable products. For the smaller number of products where only petroleum plastics will suffice for the foreseeable future, like certain medical devices, then oil can be reserved for those limited uses until practical alternatives are developed. In the end, the economy must find a way to make do with many fewer plastic products. Fortunately, public opinion supports cutting back on plastic, especially in packaging.[15] As part of a larger cultural change and if given practical alternatives, Americans will likely be willing to use much less plastic in their daily lives.

Before the Civil War, slave plantation owners and their spokesmen in Congress warned that it was impractical or even dangerous to emancipate enslaved people.

First, the only way to free slaves without destroying the economy or violating the property rights of slaveowners would be to compensate them for the fair market value of their freed slaves. And that would bust the federal budget, a claim we'll examine in more detail below.

Second, even if emancipation was possible without destroying the economy, planters claimed that allowing free Black people to live side by side with their former masters would create social tensions that would likely lead to violence. Finally, freed Black men would compete with white workers for jobs, leading to more social tension and adding to the lists of the unemployed and destitute in both races.

South Carolina's pro-slavery advocate James Henry Hammond opined that every country whose lower class was composed of free laborers had to live in constant fear of revolution and insurrection by that working class. But in the South, where half the working class were slaves and the rest were poor whites who served on slave patrols, and thus had an interest in maintaining the system, no standing armies or police forces were necessary to maintain order. "Small guards in our cities, and occasional patrols in the country, ensure us a repose and security known nowhere else."[16]

As in his other defenses of slavery, Hammond was underplaying the conflict that slavery created along with the white-on-white violence that the South was famous for in the years leading up to the Civil War. It's true that armed slave rebellions in the American South were rare compared to other slave areas in the Western Hemisphere like the Caribbean. Yet, enslaved people found many nonviolent ways to resist their enslavers, from running away to equipment sabotage to work slowdowns.

Of course, enslaved Black people themselves were subject to brutal violence and the constant threat of violence from white enslavers and overseers. And poor Southern whites experienced notoriously high levels of violence in their own communities in the first half of the nineteenth century as well, violence which later exploded on the battlefields and even in home-front communities when the Civil War broke out.

Slavery and the rigid class system it created was nothing but violence for both Black and white Southerners. But planters continued to claim that slavery was the "cornerstone" of the South's supposedly peaceful republican government, with great benefits for all classes of citizens, but especially the poor. As we'll see next, that same hollow concern for the wellbeing of the poor is also expressed today by Big Oil.

3. Poor People Will Be Hurt the Most

Shedding big crocodile tears, Big Oil and their allies claim that leaving fossil fuels in the ground will hurt poor people more than anybody else. This claim takes lots of guts to make, since it's clear to anybody who's ever seen an oil refinery, natural gas compressor station, mountaintop-removal coal mine, or coal ash dump that the people who live near dirty energy facilities suffer the most from fossil fuel production. And residents of frontline communities are nearly always low-income families and people of color. The asthma, cancer, and other health problems they suffer because of pollution from fossil fuel production result from environmental racism.

But dirty energy PR agents have claimed the exact opposite of this obvious truth. Energy Citizens, another online lobbying campaign set up by the American Petroleum Institute, says that energy from fossil fuels is not just about running lights and driving cars but is about helping the poor. "It is about lifting families out of poverty, giving greater access to advanced technologies, and creating lifesaving products that are essential to everyday lives…Over the past three decades natural gas has helped 1.3 billion people worldwide get electricity and escape debilitating poverty." The industry also touts the millions of jobs in oil and gas ranging from offshore oil workers to geologists and computer modelers.[17]

In his role as attack dog for the industry, climate science downplayer Michael Shellenberger runs in to throw shade on the favorite energy source of poor people worldwide, solar power. He tells a surprising story about villagers in India who got a free solar energy system but then rejected it in favor of dirty energy from the grid. The village of

Dharnai in the impoverished state of Bihar "made worldwide headlines after it rebelled against the solar panel and battery 'micro-grid' Greenpeace had created as a supposed model…'We want real electricity' chanted villagers at a state politician, 'not fake electricity'…By 'real electricity' they meant reliable grid electricity, which is mostly produced from coal," Shellenberger writes.[18]

The devil is in the details on this story. There was nothing wrong with the solar-powered microgrid installed in the village by Greenpeace. It would have worked fine for its intended purpose, to allow ordinary villagers to use electric lights that don't need much power. The problem was that rich people plugged in big appliances that overtaxed the system.

It's true that electricity does help raise people out of poverty in developing nations, and it's better if that electricity can be reliable. But it's silly to imply that fossil fuels are the only source, or even the best source, of affordable and reliable electricity for new places. Today, the cheapest and easiest way to bring power to run lights and laptops to a village in Africa or Asia is usually not to connect the village to a coal-powered national electrical grid. But unfortunately, that's what the Indian government decided to do in Dharnai after it turned out that the small microgrid put in by Greenpeace was not powerful enough to run the energy-inefficient refrigerators and televisions of the village's wealthiest residents, appliances that sucked up all the system's power.[19]

Nearly always, the cheapest way to bring electricity to a new area today is to put up a few solar panels on a homeowner's roof or in their front yard. "It's now cheaper to build and operate new large-scale wind or solar plants in nearly half the world than it would be to run an existing coal or gas-fired power plant," according to a 2021 report by Bloomberg New Energy Finance.[20] Intermittent solar and wind power can be made available beyond the daytime with a battery. Together, solar and battery microgrids are a reliable and affordable source of power especially suited to remote locations far from the electric grid.

With renewable energy more accessible than ever, fossil fuels are no friend of the poor. Quite the opposite. Local pollution from dirty energy facilities hurts low-income people the most, which is the biggest complaint of the environmental justice movement that resurged after the Black Lives Matter protests in the summer of 2020. Ironically and tragically, as environmental justice advocates point out, climate change also hurts poor people the most, both because they tend to live in areas more vulnerable to coastal storms and because they have less ability to pay for air conditioning and are thus more exposed to heat waves.

As to the industry claim that abolishing oil will put people out of work, a just transition as proposed in the Green New Deal will help oil and coal industry workers retrain and requalify for jobs in clean energy, energy efficiency, and clean technology. According to numerous studies, the clean energy economy will be net positive for the economy in terms of creating jobs, opportunities for entrepreneurs, lower costs for consumers, and predictable energy costs free from the swings of volatile global markets for oil, gas, and coal.[21]

What about fossil fuels helping to make consumer products affordable? It's an open secret of the consumer economy that cheap products benefit the world's poor far less than they help consumers in the richest countries, who do most of the world's driving and flying, consume most of the world's manufactured goods, and run most of the world's air conditioning and heating. As a result, just 10 percent of people on earth, those living in the richest countries, produce half of the planet's fossil fuel emissions from individual consumption. By contrast, the poorest 50 percent, about 3.5 billion people, contribute only 10 percent of climate pollution.[22]

And within the wealthiest nations, the richest people are most to blame for climate emissions. Wealthy people build the biggest houses that use the most materials and take the most energy to build and operate, they take the most flights the longest distances, they drive the largest quantity of luxury cars that prioritize performance over fuel economy, and they buy, ship, use, and throw away the most consumer products. Rich people are also the ones with the most extra income to own stocks and other investments in extractive and polluting industries like oil, gas, coal, plastics, chemicals, mining, lumber, and real estate development.

"Ecological breakdown isn't being caused by everyone equally," concludes economic anthropologist Jason Hickel. "The richest one percent emit 100 times more than the poorest half of humanity. If we are going to survive the twenty-first century, we need to distribute income and wealth more fairly."[23]

Today Big Oil claims that it will hurt poor people the most if they can't get cheap access to fossil fuels. This is reminiscent of the claim by Southern planters before the Civil War that slaves themselves, along with poor white people who didn't own any slaves, would be hurt more than anybody else if slavery were abolished.

"Slavery is said to be an evil," proslavery advocate James Henry Hammond said in a speech in 1836. "But is no evil. On the contrary, I believe it to be the greatest of all the great blessings which a kind Providence has bestowed upon our glorious region…As a class, I say it boldly; there is not a happier, more contented race upon the face of the earth…Lightly tasked, well clothed, well fed—far better than the free laborers of any country in the world…their lives and persons protected by the law, all their sufferings alleviated by the kindest and most interested care."[24]

In the free labor system of the North, so argued Hammond and other Southern leaders, workers are exploited in a system of "wage slavery" where employers are allowed to pay starvation wages but are required to take no responsibility to make sure that their workers don't starve. By contrast, slavery apologists argued that it was in the financial interest of a slaveowner to care for the health and wellbeing of his enslaved workers. Except for rare cases of negligent or cruel masters, planters claimed that slaves were fed, clothed, and housed simply but comfortably and generally not overworked. Freeing slaves would leave them without work, home, or hope in a hostile land, slaveowners argued.

The myth of the kind slaveowner and the contented slave is so easy to debunk today because history has already debunked it for us. In the decades before the Civil War, hundreds of enslaved people sought to escape bondage by running away to freedom to Northern states and Canada, many with the help of the network of Black and white abolitionists known as the Underground Railroad. These runaways would not have willingly parted from their families and put their lives at risk unless the sufferings of slavery were severe and often impossible to bear.

Self-liberated slaves like Frederick Douglass, Harriet Tubman, and Sojourner Truth testified from their own experience that slavery was a living death. And white slaveowners agreed that slavery's purpose was to convert people into money, "into machines who existed solely for the profit of their owners, to be worked as long as the owners desired, who had no rights over their bodies or loved ones, who could be mortgaged, bred, won in a bet, given as wedding presents, bequeathed to heirs, sold away from spouses or children to cover an owner's debt or to spite a rival or to settle an estate," as *New York Times* reporter and author Isabel Wilkerson explains.

Treating people as property could never be anything but cruel. Enslaved people were "regularly whipped, raped, and branded, subjected to any whim or distemper of the people who owned them. Some were castrated or endured other tortures" that later generations would have defined as war crimes.[25]

Abraham Lincoln once quipped that "Whenever I hear anyone arguing for slavery, I feel a strong impulse to see it tried on him personally." Not surprisingly, even among those who insisted that the alleged security of slavery was better than the insecurity of freedom, there's no evidence that Lincoln's offer ever found any takers.

Enslaved workers were used to supporting not just themselves but also their masters, as abolitionists pointed out. And once they did get their freedom, whether one at a time before the Civil War or *en masse* after emancipation, former slaves thrived as farmers and tradespeople but also as teachers, attorneys, and physicians. Former slaves also served as elected officials during post-war Reconstruction while federal troops occupied the South and enforced voting rights laws. But when the Army withdrew from the South at the end of Reconstruction, former Confederates took over Southern state governments again. Under these "redeemer" regimes, discrimination and violence put many freed people back into poverty and different forms of pseudo-slavery such as sharecropping, the labor debt system, and convict labor.

As to poor white people across the South, they were kept in poverty by competition from large slave planters for land and agricultural markets. But the planter elite nonetheless argued that slavery was good for the poor white family that couldn't afford any slaves themselves. No matter how low a white person sunk economically, this argument went, he or she would still possess higher social status than a Black slave.

Unfortunately, some poor white people across America have continued to cling to a version of this belief into the present day, keeping them from finding common ground with low-income people of color to fight for their common interest in a clean, fair economy with good jobs and prosperity for all regardless of race or income. And it is these same low-income Americans, of all races, who have long suffered the worst effects of pollution from fossil fuel facilities and who also stand to suffer most in the future from storms and flooding, heatwaves, and droughts along with the spread of tropical diseases brought by climate change.

To make it look like fossil fuels companies' worst victims are really their beneficiaries, the industry has recruited leaders of civil rights groups and Indian nations as spokespeople for dirty energy talking points. Oil and gas companies want to "cast themselves as allies of communities of color and defenders of their financial wellbeing," writes *Los Angeles Times* journalist Sammy Roth, in an investigation exposing front groups paid to make it look like Black, Latino, and Native Americans want fossil fuels. "The goal is to bulwark oil and gas against ambitious climate change policies by claiming the moral high ground—even as those fuels kindle a global crisis that disproportionately harms people who aren't white."[26]

Sometimes, oil and gas lobbyists have claimed that a local civil rights group opposed some clean energy policy without even contacting the group they mention. After all, it's cheaper and easier to just lie about it than it is to convince Black and brown people that clean energy is bad for them and that fossil fuel companies are in fact their true friends.

In the summer of 2020, as protesters took to the streets across the country to draw attention to police violence against Black people after the death of George Floyd, a lobbying agency called Bracewell with oil and gas companies as clients told the news media that the mayor of San Luis Obispo, California, was "getting a lot of heat" from the NAACP over a proposal to limit natural gas hookups in new buildings. But the local NAACP chapter was as surprised as anyone by this claim. "That doesn't even make sense," said Stephen Vines, president of the SLO NAACP. "We've been fighting for environmental issues for over 60 years. We support the policy."[27]

Outright lies along with donations and pressure on civil rights groups to help with fossil fuel industry PR have gotten so bad that the NAACP found it necessary to publish a guide to help local chapters spot and avoid industry attempts to manipulate them for propaganda. Published on April Fool's Day of 2019, "Fossil Fueled Foolery" highlights ten ways in which oil, gas, and coal companies attempt to fool the public, especially Black Americans.

Kathy Egland, Chair of the NAACP National Board Committee on Environmental and Climate Justice, is angry about how dirty energy companies have tried to present themselves as friends of the poor, when the truth is that harm from companies producing fossil fuels has fallen hardest on low-income people and people of color:

> One of the most duplicitous strategies of the fossil fuel industry
> is manipulating messaging which feigns concern for the welfare
> of low income and communities of color. This is a self-serving

effort to maintain their wealth. The unmitigated gall, to use as pawns the very demographics that they have caused such disproportionate harm through their polluting practices, reflects the low levels to which they will sink. Greed has no moral or ethical bounds, and we will continue to expose their foolery in seeking to deceive our communities.[28]

4. We Can Clean Up Our Act

Except on the reactionary fringe, nobody in the energy industry talks much about "clean" coal anymore. But the petroleum industry does tout how much it's spending on new technology to produce oil and gas with fewer greenhouse emissions. "Since 1990, we've invested $356 billion to improve our environmental performance. In 2017 alone, we spent $15.9 billion on new technology, cleaner fuels and other environmental initiatives," says the oil industry's Energy for Progress website, referring to technology that includes carbon capture and sequestration intended to burn fossil fuels without adding greenhouse gases to the atmosphere.[29]

Sadly, in the world of oil and gas production and pollution controls, a few hundred billion dollars just doesn't go as far as you might think. Over the same period, oil company investments in environmental technology paled before the trillions they've received in government subsidies and other forms of support paid for by taxpayers. Meanwhile, carbon capture technologies remain unproven. Even if oil companies can successfully trap carbon dioxide from a power plant burning natural gas and then pump that CO_2 into an old oil well, for example, and even if the gas doesn't find a way to leak out over the following century or two that everybody hopes it will stay down there, an operation so complicated to start and maintain will be too expensive to be practical.[30]

Maybe that's why all that industry spending on burning oil and gas more cleanly doesn't seem to have done much to reverse the overall trend of rising climate pollution. During the same 30-year period touted by the industry as an era of cleaner oil and gas, global greenhouse gas emissions doubled. You can't blame that all on the American oil industry. And it's true that the United States has fallen from the world's top emitter of greenhouse pollution to the number two spot. Now, with the growth of its economy and a buildout of coal plants,

China has become the world's largest climate polluter. But to consider Chinese emissions a solely Chinese problem is to ignore that much of that pollution is generated by factories, trains, trucks, and ships that make and transport goods to the American market. As American industrial companies have offshored their manufacturing, they've also offshored their greenhouse emissions.

To continue the analogy to abolition, in the mid-nineteenth century Southern planters who presented themselves as progressive claimed that they could make slavery more humane by bringing treatment of enslaved people up to the standards expected by a more advanced era. On the eve of the Civil War, proslavery leader James Henry Hammond even compiled a detailed manual for operating his plantation in South Carolina purporting to mandate the fair and kind treatment of enslaved workers, with rules offering special protection to pregnant women, nursing mothers, and the elderly.[31]

Yet, enslaved people were obviously not impressed by such reforms. Enslaved Black Southerners continued to vote with their feet, running away to freedom in Northern states or Canada when they could and otherwise offering myriad acts of quiet resistance to slave masters' control back home.[32]

5. We're Not to Blame—You Are

Big Oil has long tried to shift the blame for pollution, and thus the responsibility for solutions, from themselves to ordinary consumers. Major oil corporations doing business in the United States including ExxonMobil, Shell, BP, and Chevron are among the 100 companies across the globe that are responsible for 71 percent of greenhouse emissions in the last three decades.[33]

Those companies don't want you pointing the finger at them for selling oil and gas. They say that they're just meeting market demand for energy. Instead of blaming them, they want you to look in the mirror and conclude that you're the culprit for using oil and gas and for buying products and services that also use dirty energy.

To deflect blame away from itself as a major oil producer and lay that blame on the consumer is why BP, the world's sixth largest polluter, originally came up with the concept of the "carbon footprint," encouraging consumers to monitor and try to reduce our use of energy, water, and other resources.

"More than 20 years ago, one of the company's marketing campaigns helped cement the perception that the responsibility for reducing emissions lay with individuals, working the phrase 'carbon footprint' onto our tongues," as Kate Yoder writes in *Grist*. "The underlying message: Let's talk about how to solve your emissions problems." Now, BP is helping develop an app called VYVE that will let you track your carbon emissions from driving to work, to the store, or to pick up the kids after school. Then, the app will prompt you to buy carbon offsets to fund such projects as planting trees in Britain or providing cleaner cook stoves to people in Mexico.[34]

BP has also committed to reducing its own production of oil and gas in the future, while transitioning into renewable energy. The company says that the VYVE app is part of their effort to enlist consumers in the fight to cut back on using oil. It's too early to tell whether the company is serious this time or if they're just laying on a new shade of the old greenwash they've painted on their dirty operations in the past. But citizens have a right to be skeptical that an oil company will choose to sell less oil, and voluntarily cut their profits, just because it's the right thing to do. That's never what they've done in the past.

Until shut down by activists, BP was even trying to present itself on social media as a leader in helping consumers go green. "Find out your #carbonfootprint with our new calculator & share your pledge today!" the oil company tweeted in October 2019, sharing an online calculator to help consumers determine their carbon footprint. Climate activist Mary Heglar's response went viral: "Bitch what's yours???" Heglar explained how the tweet was part of a new approach to exposing greenwashing by dirty energy companies that she calls "greentrolling": "They can just walk out on the biggest arena in the world and pretend that they're something that they're not. And it's really persuasive. If I didn't know better, I would believe that BP was on the right side of history."[35]

Tired of climate-conscious people shaming each other for such offenses as eating meat or flying, Heglar wanted to put the blame where it belonged, squarely at the foot of oil companies. Every day, she made a point of monitoring and responding to the tweets of BP, Shell, ConocoPhillips, and other oil companies (Exxon already blocked her from its Twitter feed). Heglar's replies pointed out the companies' hypocrisy to help protect the public from the companies' PR spin and show the companies that someone was watching them. Apparently, it worked. For a few days, Shell, Chevron, and even Exxon stayed quiet on Twitter.

Let's spend a little more time on this excuse about the role of consumers in driving the climate crisis. Over the years, many well-intentioned environmentalists have also asked consumers to do their part through making their individual lives more sustainable.

"The environment is really the only major global crisis that we assume you can solve from your kitchen," explains historian and ecological author Jenny Price. "We don't assume you can solve the Middle East crisis, or child poverty, or the immigration problem even though your daily decisions are very bound up with those problems. We have to stop believing that we can solve it from our kitchens and start working for big system changes."[36]

Abolitionists also tried taking personal actions to harm the economy of slavery, especially boycotting sugar or cotton made with slave labor. Some abolitionists went much further by helping enslaved people escape bondage through the Underground Railroad. But abolitionist leaders knew that such personal actions, no matter how many people performed them, would never be enough to end the system of slavery.

In the environmental movement, personal conservation has a long and noble pedigree in principled calls to frugal living from religious and moral reformers. And it's certain that, whatever happens with the climate, in the future the United States and other rich nations will need to cut excessive consumerism which fools us into excessive getting and spending that doesn't make us happy while damaging the planet.

A whole library of articles and books with titles like *Affluenza: How Overconsumption is Killing Us—and How to Fight Back* and *The Overspent American: Why We Want What We Don't Need* along with documentary films like *The Story of Stuff* diagnose the problem of consumerism and offer their audiences the solution of buying less to solve numerous problems in their personal lives and in society alike.

Of course, people everywhere will be happier if they have enough stuff to meet their material needs for water, food, shelter, safety, and health along with their social needs for beauty and social acceptance. But advocates of simple living have proved that consumption beyond a basic level adds little or nothing to personal happiness while using up valuable resources and creating pollution and garbage. There's a reason why every major religion has identified greed as a sin and warned its adherents to guard against giving in to the temptation to acquire too much stuff.

Anyone who drives a car, uses electricity from the power company, buys food from the grocery store, orders books from Amazon, flies on an airplane, throws away a piece of plastic, or does any one of a thousand other things that are part of daily consumer life bears some responsibility for the climate crisis. "Just as adding a spoonful of sugar to her tea connected an eighteenth-century London matron to the bloody and deadly slave trade," writes Andrew Nikiforuk "so the purchase of gasoline today ties every driver to petro-kingdoms, poisoned waterways, and political corruption."[37]

But let's put that responsibility in perspective. Some of us bear much more responsibility than others. Most people across the globe who live in developing nations did little or nothing to pollute the atmosphere with greenhouse gases. Citizens of rich nations are each responsible for more pollution from their lifestyles. But compared to oil companies and wealthy individuals, the average middle-class or working American bears very little responsibility for climate pollution.

Despite what oil companies (and some environmentalists) say, individuals making changes to their personal lifestyles will do little to avert climate disruption, according to scientists. Michael Mann, author of the famous "hockey-stick" graph of climate pollution and analyst of the "new" war by oil companies on the climate, recognizes that we all face daily choices that impact the climate. "Do we turn the lights on in the morning, or is the light of daybreak sufficient for finding matching socks? Do we feast on bacon and eggs for breakfast, or will a bowl of oatmeal suffice?"

But changing your life to get the right answers on such questions will do little to stop runaway climate disruption, since altogether, personal choices add up to very little pollution created or saved. Even worse, Mann thinks a focus on personal virtue is distracting and counterproductive.

> There is also a lot of finger-pointing going on and, some argue, virtue signaling. But who is truly walking the climate walk? The carnivore who doesn't fly? The vegan who travels to see family abroad? If nobody is without carbon sin, who gets to cast the first lump of coal? If all climate advocates were expected to live off the grid, eating only what they could grow themselves and wearing only the clothes they'd knitted from scratch, there wouldn't be much of a climate movement.

Mann urges people who care about climate to turn their attention away from personal choices and towards politics. Practical solutions to climate are available but they require government to pass laws and make regulations to encourage or require companies to switch from products that pollute to ones that don't: "We don't need to ban cars; we need to electrify them (and we need that electricity to come from clean energy). We don't need to ban burgers; we need climate-friendly beef. To spur these changes, we need to put a price on carbon, to incentivize polluters to invest in these solutions."[38]

Focusing on consumer actions is a hard habit to break. Ordinary people have been guilt tripped for years into thinking that using and throwing away too much stuff has caused climate change and the larger ecological crisis. The average American today did not choose to live in an economy that creates so much waste and pollution. If we had the choice to live a comfortable life that was also climate-friendly, most of us would certainly do things differently. Most of us are locked into a system where we have no choice but to buy gasoline cars to commute to work doing things that are bad for people or the planet or both to pay rent or mortgages to rich people who will put their money into lavish consumption or investments that create more harm.

Any one person or household has little control over the system. Even a bunch of households working together as consumers—whether hundreds, thousands, or millions of households—isn't the answer. All the voluntary simple living in the world won't be enough to stop ice caps from melting and seas from rising, nor will personal eco-virtue be enough to tamp out runaway wildfires and cool down heat waves.

"I work in the environmental movement. I don't care if you recycle," writes Mary Heglar, who, when she's not greentrolling oil companies, works for the Natural Resources Defense Council. "Stop obsessing over your environmental 'sins.' Fight the oil and gas industry instead."[39]

Just as today's Oil Power has tried to deflect the blame for polluting the climate from companies that produce fossil fuels to the consumers who use them, so the Slave Power of the 1840s and 1850s tried to shift any blame for slavery—if they admitted any blame—away from planters who produced cotton and other commodities with enslaved labor and onto consumers who bought cloth and other goods produced from crops grown on Southern plantations. There is truth to the claim that since every white American, including Northerners, benefited from slavery that all white people were complicit in the evils of slave agriculture. For their part, Southern planters said that they were merely meeting world demand for cheap cotton and other commodity crops.

Just as environmentalists have disagreed about the role of individual conservation vs political action, so before the Civil War, abolitionists grappled with the best way to end slavery, and the proper lifestyle for white people of conscience.

Over nearly a century, abolitionists tried a variety of individual actions to chip away at slavery. These included helping enslaved people to free themselves by escaping to the North or to Canada and running boycotts or making moral appeals to pressure Southern slaveowners to voluntarily manumit their enslaved people. But despite these virtuous efforts, slavery continued to grow.

"During three long decades of struggling to change people's minds, abolitionists watched helplessly while slaveholders doubled their portion of the national domain," writes historian James Brewer Stewart. "Profits wrung from enslaved labor doubled then doubled again. The enslaved population shot up by 25% from three to four million. In response to 'moral suasion,' slavery waxed fat, white supremacy endured."[40]

Abolitionists clearly had not convinced Southern planters to give up their slaves for moral reasons. Cotton growers were so anxious to get even more slaves right up to the Civil War that they took the risk of knowingly purchasing smuggled cargoes from Africa, in contravention of a constitutional ban on the international slave trade dating back to 1808.

Frustrated at the failure of persuasion, abolitionists then tried individual action through boycotts of slave-made products. After seeing little evidence of success, abolitionists concluded that individual actions would never be enough in themselves to end American slavery. Individual actions by themselves were a drop in the bucket. To make any difference, personal lifestyle changes needed to be part of a campaign. And that campaign had to go beyond persuading the public that slavery was evil so that more people would voluntarily free slaves. The campaign had to demand that government compel slaveowners to give up their property in humans. This meant campaigning for legally mandated abolition at the federal level that would override state laws and apply nationwide.

This was a radical idea at a time when the federal government was small and it took the armed revolution in American law brought by the Civil War to make national action against slavery acceptable politically.

Once the cannons started roaring, Abraham Lincoln was able to put into motion his abolition plans with the justification that they would help the Union win the war.

First, in 1863 he issued the Emancipation Proclamation to free people enslaved in any areas controlled by armed insurrectionists (meaning Confederates). Two years later, Lincoln allied with Radical Republicans in Congress to pass the Thirteenth Amendment ending slavery across the country. Unfortunately, the amendment contained an exception for those convicted of a crime, an exception that would allow forced labor to continue in prisons, on prison farms, on chain gangs, and even at privately owned farms and businesses through convict leasing programs. Nonetheless, the Thirteenth Amendment remained the heaviest blow ever dealt to slavery in world history, freeing more people at one stroke than ever before.

It's exactly this scale of oil abolition nationwide that is required to cut America's fuel pollution enough to make a difference on climate in the next ten years. When it came to slavery, it took a war that killed more than 620,000 Americans to finally end it. Later, we'll discuss whether such an abolition of oil will be possible today without the equivalent of war.

As to assigning blame for slavery in the nineteenth century, there was plenty to go around. But the big planters deserved most of the blame, because they were the ones who did the most damage both to enslaved people and to American democracy.

In the decades leading up to the Civil War, every free American in the North, the South, or new areas in the West who wore clothing made from cotton cloth put money in the pockets of cotton planters in Mississippi or Alabama. In turn, those planters used earnings from selling crops to buy more enslaved "hands" and more land to put those hands to work planting, tending, and harvesting. Likewise, whether a man in Chicago smoked a pipe stuffed with tobacco from tidewater Virginia or a woman in Philadelphia cooked a meal with rice from the soggy lowlands of South Carolina, consumers of products besides cotton also benefitted from the original source of always-low prices in the slave agriculture industry.

Finally, blame for slavery also accrued to workers in New England clothing mills who depended for their jobs on a ready supply of raw material from the Cotton Kingdom of the Southern states or the dock workers in New York City who made their living unloading cotton bales off ships from New Orleans or Mobile.

In their roles as both consumers and workers, ordinary white people in free states were fed and clothed by slaves just as white Americans in the South were. Likewise, many Northerners' jobs depended on the ability to process raw materials produced by enslaved Southerners. Even the tax system spread the guilt for Southern slavery across the North. Henry David Thoreau spent two years at Walden Pond partly to protest the role those Northern taxpayers were forced to play in subsidizing the Mexican War that would open new lands to slaveholders.

In the South, white people of all classes played an even bigger role in supporting slavery. Just before the Civil War, one in three white families across the states of the future Confederacy owned at least one slave. For Southern white people, buying a slave was like getting a BMW today—it established its owner as a member of the elite. So, those whites who didn't yet own any slaves aspired to join the slave-owning class if they had any prospect of doing so. Surprisingly, the huge mass of poor whites who knew they'd never rise from the lowest class to own any land or any slaves at all also supported slavery by serving on slave patrols and voting for politicians who reminded poor white people that, no matter how low their fortunes might sink, at least they'd never be Black.

Just as most of the blame for climate pollution today goes to the 100 companies that produce more than 70 percent of greenhouse gas emissions, so most of the blame for slavery in the nineteenth century went to just a few big Southern planters. The biggest producers of cotton owned most of the slaves, most of the land, and controlled politics not just in their region but on the national level. Of the first eighteen presidents of the United States, at least twelve owned slaves.[41] Slaveowners also dominated Congress and the Supreme Court. Most of these were wealthy men.

Though slave-made products were found across the economy, only a small group of elite Americans were responsible for keeping most enslaved people in chains. In 1860, a year before the Civil War broke out, less than one percent of white Southerners owned a quarter of all slaves.[42] Those rich slaveholders also did most of the lobbying to fight off government actions to contain slavery's spread. Those elites also spread most of the propaganda claiming that slavery was necessary to the economy and beneficial to master and slave alike.

What Lincoln said about the average Southerner in 1854 can apply well to the average American today. "Let me say that I think I have no prejudice against the Southern people. They are just what we would be in their situation. If slavery did not now exist amongst them, they would not introduce it. If it did now exist amongst us, we should not instantly give it up—this I believe of the masses north and south."[43]

Southerners didn't invent their monstrous peculiar institution. They inherited slavery from Europeans who brought bound workers to America in the colonial era before the United States was an independent nation. In the same way, today's American had no role in creating an industrial economy based on fossil fuels.

To succeed at abolishing oil and saving the climate, we'll need to lay the blame for the problem of dirty energy, and the responsibility to make the most sacrifices for climate solutions, squarely where that blame and responsibility belongs. It's not on the average consumer who drives a gasoline car because she has little other practical alternative. It's on dirty energy companies themselves who created a system that gives that driver so little alternative and continue to pull political dirty tricks to protect a system that's become oppressive and dangerous to all Americans.

6. Your Solution is Too Expensive and Leads to Tyranny

Finally, Big Oil and their allies say that shifting America to 100 percent clean energy fast enough to cut greenhouse emissions in half by 2030, as proposed by the Green New Deal, will be too expensive and run up too much debt for the federal government. The Senate Republican Policy Committee has ridiculed the Green New Deal as a "crazy, expensive mess."[44] Even worse, critics warn, expanding government enough to administer all the programs required to build out clean energy across the country will risk turning America into a "socialist" dictatorship.

For years, conservatives have complained that the federal government carries too much debt and that we can't afford to invest in programs to replace fossil fuels such as incentives for solar and wind power. And those complaints have gotten louder after the pandemic shutdown's effect on the economy. The federal government spent a record amount of money on Covid relief programs while tax revenues were down because businesses were closing and workers were losing their jobs, either temporarily or permanently. There's no question that public resources were stretched thinner than they have been in a long time.

But even during the depths of the Great Depression, FDR found enough money to start the New Deal, investing in the economy and in communities and putting millions of people back to work. And just as the Civil War did, World War II put even more people to work, leading to another era of prosperity after the war. Yet, back then, Republicans complained about socialism too. The definition of the word "socialism" seems to be pretty much whatever the speaker wants it to be, and free-market PR guys throw around the term to try to scare the public off from regulating big business.

But one thing history tells us for sure. Government action that got Americans together to fix the economy or fight Hitler didn't create a dictatorship back when FDR was president. So, it's fair to predict that government action to mobilize the economy to fight the climate crisis won't destroy the American republic now either.

The fact is that today there's plenty of money to transition to clean energy and clean up other pollution. The Green New Deal will cost much less per year than it cost for coronavirus relief in 2020, and the Green New Deal will be a better public investment than Covid relief, which went mostly to dirty energy interests and other big corporations but did very little for ordinary Americans. If just 12 percent of stimulus funding that governments around the world have already promised for Covid were spent every year through 2024 on low-carbon energy investments and reducing our dependence on fossil fuels, claim researchers in *Science*, that would be enough to limit global warming to 1.5 degrees Celsius, the most ambitious target in the Paris Agreement.[45]

"One of the messages of this is that tackling climate change is not that expensive, which is contrary to the narrative that we often hear— that this is a crazy, expensive mess," writes Marina Andrijevic, one of the paper's authors.[46]

Cutting subsidies to fossil fuels, increasing the energy efficiency of federal government operations, and other measures to cut fossil fuel costs could be combined with changes to the tax code intended to make rich people and big corporations pay their fair share and "would provide trillions of dollars to the government over the next ten years, money that could be spent to fight climate change," writes economist Joseph Stiglitz. To sweeten the deal, helping businesses and homes to use energy more efficiently while creating millions of jobs would create a powerful economic stimulus. "I suspect that between the stimulus to growth that the GND would provide...and the redeployment of resources and tax measures...there would be more than sufficient resources to win the climate war," explains Stiglitz.[47]

Big Oil and their confederates in the free-market lobby also claim that the Green New Deal would threaten the American way of life by bringing "socialism" and "big government." This laissez-faire argument was refined in the anti-Communism of the Cold War. Before that, the sanctity of the free market unhindered by government oversight was offered by robber barons to fight labor regulations during the Gilded Age.

But if you go further back, this exact same argument can be traced directly back to the proslavery ideology of the antebellum South. Planters claimed that abolition would trample on the right of Southern states to make their own decisions about slavery free from outside interference. Prominent proslavery writers argued that abolition would bring socialism, feminism, and many other fearful "-isms" along with abolitionism that would threaten American freedoms.[48]

You couldn't make war on King Cotton, after all, because if you tried, you'd be hurting yourself. The interest of a few rich cotton planters in Southern states was identical with the public interest across the country—at least that's what the planters' propaganda said.

Near the beginning of the Civil War, Abraham Lincoln came up with a plan for compensated emancipation in border states that remained in the Union but still had slaves. Lincoln proposed that the federal government should pay $400 per slave to slaveowners in Delaware, Maryland, Missouri, Kentucky, as well as the District of Columbia. This would add up to $173 million, which would be the same amount of money that the federal government would spend over 87 days of fighting the war, a war that Lincoln hoped to end by freeing the slaves. Under this plan, emancipation would not be immediate, but would take place gradually over several decades, with the last slaves slated to be freed in the year 1900.

Today, a plan like this sounds preposterously slow and underpowered. And why should guilty slaveowners be compensated for ceasing to hold people in bondage? The abolitionist minority raised such questions at the time, but in mainstream politics such objections never arose. Instead, border state congressmen overwhelmingly rejected the president's plan as too radical—and too expensive, as they told Lincoln in a letter. "We did not feel that we should be justified in voting for a measure, which, if carried out, would add this vast amount to our public debt, at a moment when the Treasury was reeling under the enormous expenditures of the war."[49]

As it turned out, creative financing by Lincoln's Treasury Secretary Salmon P. Chase brought in funding to fight the war for not just 87 days but for three more years. Afterwards, though the economy of the South was devastated by war and disrupted by emancipation, the U.S. economy enjoyed decades of boom times in the post-war Gilded Age.

The border state congressmen responded to Lincoln's proposal that letting the federal government purchase their slaves would set a precedent for national interference in affairs of state governments, "a radical change in our social system" that would lead to tyranny and an eventual overthrow of the republican system of government framed in the Constitution. "The right to hold slaves is a right appertaining to all the States of this Union. They have the right to cherish or abolish the institution as their tastes or their interests may prompt, and no one is authorized to question the right, or limit its enjoyment."

Leave us alone, these slave states were saying to Lincoln. We'll abolish slavery when we're good and ready, if we ever are ready. If Lincoln had deferred to such an interpretation of freedom for states, then freedom for slaves may well have had to wait until after the Civil War or even after the beginning of the twentieth century as proposed in Lincoln's gradual, compensated emancipation plan.

We face a similar situation today. If Americans wait to abolish oil until the oil industry is good and ready to give up their property in oil reserves and their prospect of future profits selling us oil, gas, and feedstocks for plastics, then we may have to wait until long past the deadline to stop climate disaster. For one industry to claim that its right to an unimpeded free market to sell dirty energy trumps the right of all Americans (and everybody else on earth) to a livable climate, then that's just using the word "freedom" to stand in for a new kind of oppression.

4

Climate War Worse than the Civil War

Humanity is waging war on nature. This is suicidal. Nature always
strikes back—and it is already doing so with growing force and fury.

—U.N. Secretary-General António Guterres[1]

There is a possibility of wars, even nuclear wars, if temperatures rise
two to three degrees Celsius.

—Gwynne Dyer, global security analyst[2]

Aggressors often claim that when they start a fight, they're just
defending themselves. It always turns out to be a lie. When
Confederates fired on Fort Sumter in Charleston Harbor in
April 1861, thus letting slip the dogs of Civil War, they claimed that
they were defending their freedom. But the main right that concerned
their states, to keep slavery, wasn't even at issue in 1861. Though
Abraham Lincoln was indeed America's first antislavery president, he
was only elected to stop slavery's spread to new areas. Lincoln was not
elected to abolish slavery where it already existed. Of course,
abolitionists wanted more, but the average Northern voter wasn't ready
to end slavery. Not yet. That may sound bad to us today, but it was a
reality of American politics in the mid-nineteenth century.

Voters wanted Lincoln to contain slavery, not end it. And when
Southern states started to secede, Lincoln's voters wanted him to stop
them from leaving and save the Union. Lincoln's first job in the crisis
was to keep the country together. To ease the fears of Southern leaders,

just six weeks earlier, in his first inaugural address, Lincoln promised not to interfere with the institution of slavery where it existed and even to suspend the operations of the federal government in the seven Southern states that had already seceded. Peace talks were even started. "In your hand, my fellow countrymen, and not in mine, is the momentous issue of civil war," Lincoln said. "The government will not assail you. You can have no conflict without being yourselves the aggressors."[3]

During the months after Lincoln's election but before hostilities broke out, to try to prevent Southern states from going any further towards starting their own nation, Congress quickly approved the Crittenden Compromise that would forever protect slavery from federal abolition wherever it existed and extend slavery along the so-called Missouri Compromise line of 1820 all the way to the Pacific Coast.[4]

Again, abolitionists weren't happy about it, but slavery wasn't threatened where it already existed before the Civil War. The truth was that Southern states voted to secede from the United States not to protect slavery but instead to keep expanding slavery in the future into new territory acquired by the federal government. They insisted that their sectional interest become the national interest, and that the private interest of big cotton planters was the same as the public interest of the American people then and into the future. That's how it had been for most of American history, at least according to cotton state leaders.

But Lincoln's election changed everything. It now seemed that the federal government would flip from decades of promoting slavery to a new attitude of discouraging slavery. If that was going to happen, then Southern planters had no further use for the American Union. That's why they were ready to start their own nation, the Confederacy, dedicated not just to maintaining but to growing slavery.

A century and a half later, fossil fuel companies and the free-market PR men who defend dirty energy claimed that they're defending the economy as well, or what they sometimes grandly referred to as the American way of life. It turns out to be just another lie, claiming that their private interest is the same as the public interest. It never was, and it isn't today.

U.N. Secretary-General Guterres has it right: The Oil Power is the aggressor and they're really making war on nature, which includes you, me, and the economy too. While claiming that they're defending us, fossil fuel companies are metaphorically firing their cannons right into the faces of the peoples of the earth. Climatologist Michael Mann agrees that the fight begun by the fossil fuel industry on the climate, though it's political rather than military, still brings deadly consequences. "We are in a war—though not of our own choosing—and our children represent unacceptable collateral damage. That is why we must fight back—with knowledge, passion, and an unyielding demand for change."[5]

The war on the earth and its inhabitants by fossil fuel corporations right now might feel metaphorical. But in the future, if fossil fuel companies are allowed to keep doing things their way, the war will become real, with actual soldiers firing real artillery in climate-ravaged places across the globe. That's started happening already in places like Syria and Yemen.

Scenario: Climate War Goes Nuclear

In the quote at the start of this chapter, defense analyst Gwynne Dyer envisioned a worst-case scenario of nuclear war. Does that sound alarmist and implausible?

Consider a story from the start of the Civil War. In the first few months after Fort Sumter fell, both sides assumed that the conflict would be over relatively quickly, in a few months at most. Few Americans in 1861 could imagine it would take four years of fighting and dying to bring the conflict to an end. And even fewer could imagine the scale of death or devastation that war would bring.

A couple months after Fort Sumter, on a sunny July day in 1861, dozens of Washingtonians, including journalists, several members of Congress, and photographer Matthew Brady packed their sandwiches, opera glasses, and families into their carriages. Then they rode 30 miles west to a place in Virginia called Manassas Junction to watch what they expected to be a splendid entertainment—trained U.S. troops quickly scattering amateur rebel forces.

But the holiday picnic soon turned into a panicked rout. After a promising start for the federal forces, Confederate troops under Stonewall Jackson launched a counterattack that sent both federal troops and civilian onlookers tumbling back to the safety of the capital city. Soon after the battle that came to be known in the North as Bull

Run and in the South as Manassas, people in both regions would learn that things would turn out so much worse than they'd thought.[6]

Underestimating the threat of war has left the United States vulnerable from the Civil War to World War II. Today, we've spent enough time underestimating the effect of climate heating on nature and civilization alike. As one scientific model after another for predicting climate damage has turned out to be optimistic, underestimating the actual damage from greenhouse pollution to the climate, the time for worrying about climate "alarmism" has passed.

We must now be willing to imagine a worst-case scenario so that we can prevent it. As climatologist Michael Mann has written, we need both urgency and agency. "The climate crisis is very real. But it is *not* unsolvable. And it is *not* too late to act."[7] And that applies as much to ice caps melting as it does to political instability and armed conflict.

Gwynne Dyer is not the only defense expert to imagine how one of the world's nine nuclear-armed nations, most likely one with a weak government located in a dry part of the world like Pakistan, could find itself hit by a weather disaster with political consequences that cascade out of control into unplanned use of the unstable country's ultimate weapons.

One scenario goes like this. Climate heating contributes to a series of heatwaves and droughts, perhaps two or three years in a row, causing wheat harvests to fail. Since the government lacks funds to purchase enough grain on a world market hit by higher prices because of droughts elsewhere, food supplies dwindle, and prices skyrocket until grain cannot be had at any price. Hungry citizens stage protests that explode into food riots in major cities which spread across the country bringing widespread civil unrest. Authorities respond with police repression, beatings, and arrests, which further enflame the population.

Military officers, some of the most trusted leaders in a nation whose civilian government is infamous for its corruption, step in to stop the violence. Generals and colonels depose the flailing president in a coup and form a national emergency government.

To quell popular unrest and solve the water crisis, the new junta tries to get access to water resources fast by demanding more from rivers that cross over from a neighboring nation, also nuclear-armed. A series of diplomatic threats, buildups of troops and weapons along the border, terrorist attacks inside the neighbor's territory, and finally, armed clashes between the two armies spiral out of control. As chaos strikes both governments, somebody on one side or the other fires the first missile at an enemy air base, triggering an escalated response on both sides hitting major cities and each national capital. Thus begins, and ends, the world's first, and one must hope, last nuclear war.

The good news is that such a nightmare scenario is on the unlikely end of the spectrum of climate threats to global security spun out by defense planners. The bad news is that less apocalyptic but still deadly scenarios of climate-driven conflict are much more likely soon. The worst news is that some climate wars have already started.

Lessons from America's Bloodiest War

As we consider grim pictures of armed conflict driven by or at least made worse by global heating, it's helpful to remember what else U.N. Secretary Guterres said in his 2020 State of the Planet address. "Human activities are at the root of our descent toward chaos. But that means human action can help to solve it. Making peace with nature is the defining task of the twenty-first century. It must be the top, top priority for everyone, everywhere."[8]

In the meantime, our cold war on the planet threatens to erupt into hot war among nations and armed insurgent groups. An army of experts on global conflict, from seasoned defense analysts to retired generals and admirals to sitting heads of state, agree that global heating has already helped to inflame armed conflicts across the globe from the uprisings of the Arab Spring to the Syrian civil war.

These military analysts also predict that more climate heating in the future will make future civil unrest and warfare more deadly. Along with deaths and injuries directly arising from weather disasters like heat waves, wildfires, and coastal storms, humans fighting each other for dwindling supplies of water, food, and arable land in a heating world could bring an endless future of climate wars deadlier than even the biggest armed contests of the past.

By far, the deadliest war in American history was the Civil War. From 1861-1865, more than 620,000 soldiers and sailors from both sides died from combat, accident, starvation, and disease. "The human cost of the Civil War was beyond anybody's expectations. The young nation experienced bloodshed of a magnitude that has not been equaled since by any other American conflict," according to the American Battlefield Trust. The war killed two percent of the U.S.

population at the time, which is the equivalent of six million Americans today.[9]

Nationwide, one in four men who marched off to war never returned. On the Union side, casualties were especially high among Black soldiers in units of U.S. Colored Troops, who were special targets for an enemy enraged by the sight of a Black man in a military uniform holding a gun. On the Confederate side, one in three white households in the South lost at least one family member. After Appomattox, one in thirteen fighting men came home missing at least one limb, preventing a return to pre-war work on farms or in factories.

Despite the loss of much of a generation of young men, the economy of the North enjoyed a long wartime boom that history knows as the Gilded Age. But the South was devastated by the greater loss of life and the massive damage to its farms and cities. While some Southern cities like Atlanta shared in the national economic expansion, it took a century for the South's overall economy to fully recover from the war. Compared to other regions of the country, the South still scores low on measures of prosperity like household income and education.[10]

The Civil War was the last time that large-scale conflict was fought on American soil. It may be difficult for most of us to imagine the suffering that war can bring to one's own country today, but weather disasters stoked by climate heating have already brought wartime levels of destruction to thousands of communities nationwide from Alaska to the Gulf Coast. Yet, it's deeply embedded in contemporary American culture to assume that since the last century and a half has brought wars only overseas, that the future is likely to continue to spare our shores from armed conflict.

Military planners agree that the prospect of war at home is extremely unlikely in the foreseeable future. But as decades of global heating put stress on political order across the globe, experts predict that the potential for civil unrest and even foreign invasion of the United States will increase. In the meantime, it's clear that wars in other countries will continue and that the impacts of weather disasters will make foreign wars worse. And these wars will indirectly impact the U.S. by sending waves of refugees, disrupting trade in food and other staples of the global economy, and pulling the federal government into ever more dangerous and urgent foreign entanglements.

Drawn with the Lash and Paid by the Sword

Defense planners grappling with the threat of climate disruption to world peace suggest a truly apocalyptic question: Could all the economic progress made possible by fossil fuels since the beginning of the Industrial Revolution be destroyed by armed conflict enflamed by climate chaos in coming decades?

Abraham Lincoln asked much the same question about America's economic assets in the famous speech he gave at his second inauguration in March 1865. As the Civil War was winding down and Union victory was in sight, Lincoln looked forward to peace. "Fondly do we hope—fervently do we pray—that this mighty scourge of war may speedily pass away." But he also worried that if the war continued to drag on, the cost in blood and treasure would be immense.

> Yet, if God wills that it continue, until all the wealth piled by the bond-man's two hundred and fifty years of unrequited toil shall be sunk, and until every drop of blood drawn with the lash, shall be paid by another drawn with the sword, as was said three thousand years ago, so still it must be said 'the judgments of the Lord, are true and righteous altogether.'[11]

In much the same way today, we might wonder if it's possible that all the gains that fossil fuels provided to industrial development for the last two and a half centuries could be incinerated by a global conflagration fueled by the pollution from burning those same fossil fuels. Defense planners offer several scenarios where armed conflict stoked by climate impacts will pose critical threats to the nations of the earth. We'll examine some of these scenarios below. But first, on the topic of war and climate, we must answer another question.

Can the huge project of taking apart a century or more of fossil-fuel infrastructure and replacing it with clean energy in the next three decades be done without the political equivalent of war on dirty energy companies? The world must act to save the climate much bigger and faster than we have so far. Facing both daunting logistical challenges and dogged political resistance from the Oil Power and their allies in right-wing political circles, can the war to end oil be a peaceful one?

Abolitionists had to ask themselves the same question before the Civil War provided an answer that nobody could argue with. By the

1860s, abolitionists had been campaigning to end slavery on and off for nearly a century.

Peaceful protesting and their "moral suasion" appeal to the conscience of Americans North and South had indeed turned the hearts of some white Americans towards the fight for freedom for the enslaved. But this moralistic approach was polarizing, and it also hardened other hearts to defend the white South's traditional source of wealth as a matter of pride—whether expressed as states' rights, misplaced humanitarianism, or even as personal freedom (for the slaveowner, not the slave of course).

In the end, peaceful protest by abolitionists proved insufficient, and it would take the bloodiest war in American history to end slavery in 1865.

"War simplifies things. It simplifies things politically," explains Civil War historian H.W. Brands. "Slavery was an insoluble problem for the American political system in peacetime...It couldn't resolve that problem in seven decades of peace. But then, in two years of war," slavery was first partially ended with Lincoln's Emancipation Proclamation in 1863 and then totally abolished by the Thirteenth Amendment in 1865.[12]

As Frederick Douglass explained years after the Civil War, Black freedom had not come from "sober dictates of wisdom, or from any normal condition of things." Emancipation, which Douglass and other abolitionists considered to be a second founding of the United States, "came across fields of smoke and fire strewn with...bleeding and dying men. Not from the Heaven of Peace amid the morning stars, but from the hell of war."[13]

After more than half a million dead, billions of dollars in property destroyed, and the collapse of King Cotton, it would be easy to conclude as Abraham Lincoln suggested that the Civil War did in fact destroy centuries worth of national wealth built up by enslaved laborers.

We should apply the sentiment of Lincoln at the end of the Civil War to saving the climate today: We should fondly hope and fervently pray that it will not take a similar level of violence or even an actual shooting war of some kind to end the widespread use of oil and save the climate. After the insurrection at the U.S. Capitol on January 6, 2021, it's not hard to imagine that polarization in American politics will continue to spill over from heated tweets to physical violence. Yet all Americans of good faith must work against domestic terrorism and political violence to save the climate, ensure racial equity, and solve the other big problems that our country can no longer afford to ignore.

The Civil War was so deadly because it was so well organized. After Lincoln was elected president in November 1860, one after another, Southern state governments announced their intention to leave the Union. This began just over a month after the election, on December 20, 1860, when South Carolina announced that its union with the United States was "immediately dissolved." In the weeks to come, other Southern states followed with their own secession ordinances.

Then, those states got together and started their own country, the Confederacy, complete with president and congress, flag and national anthem, and generals and soldiers in uniforms who marched to fight and die on battlefields from Manassas to Gettysburg and in cities from Atlanta to Richmond.

Whatever sectional and political differences still divide us, America just isn't that kind of country anymore. At least not yet.

But some experts think that Americans might be on the way to a level of polarization over politics not seen since 1860. "It's time for Americans to wake up to a fundamental reality," writes David French in *Divided We Fall: America's Secession Threat and How to Restore our Nation.* "The continued unity of the United States of America cannot be guaranteed. At this moment in history, there is not a single important cultural, religious, political, or social force that is pulling Americans together more than it is pushing us apart. We cannot assume that a continent-sized, multi-ethnic, multi-faith democracy can remain united forever, and it will not remain united if our political class cannot and will not adapt to an increasingly diverse and divided American public."

Not only do Americans lack a common popular culture anymore, French contends, but we also live increasingly in separate types of communities from people who don't share our tastes in music, television, or sports; who don't vote the way we do; and who don't share our religious beliefs. We can add that Americans also disagree on how to interpret the history of the Civil War and abolition as we disagree on the threat of climate catastrophe and the need for the Green New Deal.

Until party leaders stop fomenting polarization, the unthinkable becomes possible in French's analysis. "Given this reality, why should we presume that our nation is immune from the same cultural and historical forces that have caused disunion in this nation before and in other nations countless times?"[14]

Even worse, it's not enough to simply express disagreement with a supporter of a different political party anymore. Now, Americans increasingly hold people outside their political tribe in contempt and believe that if they came to power, it would destroy the country. That justifies extreme measures to keep the other side down.

Just before the election in fall 2020, one in six people, double the number three years earlier, said in a survey that violence "could be justified to advance their parties' political goals." We saw where that attitude led after Donald Trump lost the election and some of his supporters stormed the U.S. Capitol to try to stop Congress from certifying the election of Joe Biden. Writing before the January insurrection, Larry Diamond, the Stanford professor who conducted the opinion research, was concerned not just about the increasing openness to political violence but also the actual violence seen in American politics in recent years, from white supremacists clashing with anti-fascist protesters in Charlottesville in 2017 to President Trump encouraging right-wing extremists such as the Proud Boys.

"The level of armaments that these people have, the stockpiles of military-style weaponry and body armor, the high-volume gun clips—there's no precedent in American history for this, and that's why I think the current era is more dangerous than anything we've seen in decades," Diamond writes.[15]

Diamond is correct that the potential for political violence today is higher than it's been in decades, since the 1960s at least. He's also right that the weapons amassed by militia groups are higher tech than ever. But he's wrong that civilians building private arsenals and preparing to fight their political opponents is unprecedented. Before the Civil War, paramilitary groups, often growing out of political parties, formed in both the North and South. Before war broke out, such groups held marches and rallies to intimidate their political opponents.

The Wide Awakes were a youth club formed by Republicans to support Lincoln's campaign for president in 1860 with chapters across the North from California to Maine and even in slave states in the Upper South like Virginia and Missouri. They carried banners depicting their symbol, a single open eye. Marching in formation to the beat of military marches, Wide Awakes wore paramilitary uniforms of glazed-cloth caps along with dramatic capes to protect them from being scalded by hot wax dripping from their six-foot torches. Wide Awakes created a spectacle by marching in campaign parades in Boston, Cleveland, Chicago, and New York.

"Though it was a Democratic city, tens of thousands of [Republican] men in uniforms went goose-stepping down Broadway," according to historian Richard Kreitner, "cheered by spectators from windows and rooftops along the route—'a surging sea of excited humanity,' one observer recorded."[16]

The group also provided security for speakers at Republican campaign rallies, which meant they had to get a little rough when rowdies from the other side tried to make trouble. "The Wide Awakes offered a spectacle unlike anything in American politics to that point. Their menacing, martial bearing made it easier for citizens in both sections to imagine their longstanding political conflict suddenly turning into a military one," writes Kreitner.

Though Wide Awakes never marched in the Deep South, white Southerners viewed the group as an antislavery intimidation force and responded by recruiting young men into their own paramilitary group, the Minutemen. They were meant "to form an armed body of men...whose duty is to arm, equip and drill, and be ready for any emergency that may arise in the present perilous position of Southern States," according to one Southern newspaper. After Confederates fired on Fort Sumter, members of paramilitary groups enlisted in both armies. This is only one way that we know from history how citizens forming armed groups can increase political tensions and help precipitate war.

In the twenty-first century, at least our politicians haven't yet started coming to blows over politics like they used to in the years leading up to the Civil War, even inside the halls of Congress.

Today hot words may fly in the Senate between Democrats and Republicans or over Twitter. But in 1856 one of South Carolina's congressmen, Preston Brooks, entered the Senate chamber, walked over to antislavery Senator Charles Sumner at his seat, and proceeded to beat the Massachusetts senator with his wooden cane. Brooks was fined $300 and had to resign his seat in Congress. But to white Southerners Brooks became a hero, their defender against what they perceived as Northern aggression against the whole Southern social and economic order. The South Carolinian went home to accolades from white people of all classes in his home state. Over the next few months, gifts of new canes arrived at the Brooks home from across the South.

The caning of Sumner wasn't an isolated incident. In the years before the Civil War, political fights over slavery often led to fisticuffs in both chambers of Congress: "Punching. Pistols. Bowie knives. Congressmen brawling in bunches while colleagues stood on chairs to get a good look," were such regular occurrences that British diplomats concluded that the floor of the House of Representatives was too dangerous to visit.[17]

Abolitionists made a hero of Sumner and his bloodied coat took on the status of a holy relic in Northern antislavery circles. In general, abolitionist leaders were pacifists, but some urged a form of resistance that veered close to violence. Henry Highland Garnet, who escaped slavery as a child when eleven members of his family fled Maryland, called for the enslaved people of America to start a general strike.

"Cease to labor for tyrants who will not renumerate you," Garnet told a national Black convention in Buffalo in 1843. "To such DEGRADATION IT IS SINFUL IN THE EXTREME FOR YOU TO MAKE VOLUNTARY SUBMISSION...RATHER DIE FREEMEN, THAN LIVE TO BE SLAVES...LIBERTY OR DEATH...RESISTANCE! RESISTANCE! RESISTANCE!"[18]

But a century before Gandhi and Martin Luther King made nonviolence the standard of progressive activism around the world, most white abolitionists in the first half of the nineteenth century hoped to free slaves through peaceful activism and lobbying government.

After learning in August 1831 that a visionary enslaved preacher named Nat Turner had just led a slave revolt in Southampton County, Virginia that left seventy white people dead, William Lloyd Garrison wrote that he was "horror struck." "I deny the right of any people to *fight* for liberty, and so far am a Quaker in principle. I do not justify the slaves in their rebellion; yet I do not condemn *them* and applaud similar conduct in *white men*."[19]

Likewise, while expressing deep grievances against the fossil fuel industry, today's oil abolitionists remain committed to non-violent activism, as espoused by every major social and political movement in American history from women's rights to civil rights to environmentalism. People may type testy comments and post scary videos on Facebook or Twitter. But in real life, it's almost as unthinkable to imagine suburban moms in the Sierra Club taking up arms against burly Appalachian miners recruited to Friends of Coal as it is to imagine America splitting along some modern Mason-Dixon line over even the hottest issues like immigration, abortion, or gun control, not to mention climate.

Fortunately, the prospect of armed conflict seems far away from the American homeland today because war really is hell. But some of the biggest changes in American society and the economy in the past happened during wartime. Even the most committed pacifist must acknowledge that some of those changes were positive. Amid widespread suffering and death came many advances in new technology and civil rights. The best example of the latter was Lincoln's Emancipation Proclamation of 1863, which was able to free millions of slaves over the objections of many white people in the North by using the excuse that confiscating enemy "property" would aid the Union war effort.

Recently, climate activists committed to nonviolence have started to look back on those periods in history when change was speeded up by war, whether World War II or the abolition era.

In 2020 the Sunrise Movement of young climate activists, in conjunction with the For Freedoms campaign of activist artists, started a campaign called the Wide Awakes inspired by the uniformed political clubs for young men recruited by Lincoln's presidential campaign in 1860. That year the group held massive political rallies in Northern cities including a march in New York at the beginning of October.

Riffing off the old symbol of the open eye, today's Wide Awakes take an unconventional approach to climate activism, even trying to start a new holiday complete with an old-fashioned political parade to encourage the playful activism that they call "civic joy." "On October 3, we celebrated Wide Awakes Day in commemoration of the 160th Anniversary of the largest Emancipation March in American history," wrote artist and For Freedoms Founder Hank Willis Thomas.

> There are no words that could truly describe the miracles of marching down Broadway with hundreds of people dancing and singing and shouting in unison while wearing capes and ushered by Moko Jumbie stilt walkers. Or the magic of waving flags and performing music on the steps of the first site of the American Government. Collectively we claimed a new holiday dedicated to the activation of Civic Joy.[20]

Like their Civil War-era predecessors, today's Wide Awakes also push the bounds of nonviolence, at least when it comes to climate activism. Wearing Sunrise T-shirts and carrying signs that say things like "Wake Up Mitch" and "No Justice No Sleep," by cover of night they travel to homes of politicians like Senate Republican Leader Mitch McConnell who have blocked climate action. Then, standing out front of the legislator's home, they bang pots and pans and shout slogans.

Such tactics might make gentler souls uncomfortable. But Sunrise justifies making unjoyful sounds to disturb the sleep of unfriendly elected officials as an appropriate response to the climate emergency, inspired by the young political activists who made a lot of noise in the 1860 presidential campaign:

> The original Wide Awakes were relentless. And they saw results. In the next half decade, Abraham Lincoln was elected, slavery was abolished, and the visionary Reconstruction began in the South. Like our predecessors, we will demand the future we need, not the future our political leaders think they can negotiate for. Our generation will fight for true abolition and complete the unfinished work of the Reconstruction.[21]

It's refreshing to see climate activists, especially young people, take inspiration from successful and heroic movements of the American past. They clearly recognize the power of shared history to motivate climate activists today. Yet, any U.S. history buff will note how quickly the Sunrise story jumps from Lincoln's election straight to abolition and Reconstruction, leaving out the biggest part of the story: The Civil War. Of course, it's clear that the Sunrise Wide Awakes know their history, so they must be doing this on purpose. It seems that Sunrise wants to achieve the massive success of abolition through entirely non-violent activism.

Let's hope that Sunrise is correct that today's climate movement can overcome the massive power of Big Oil, a political and financial behemoth comparable to the Slave Power before abolition, without having to wage anything like a modern version of the Civil War. However, if you're looking for a precedent in slavery abolition for oil abolition, that's not how it happened in the past. Abolition came "not from the Heaven of Peace amid the morning stars, but from the hell of war," as Frederick Douglass had to remind civil rights activists nearly two decades after the Civil War.[22]

The success of abolition teaches many useful lessons for the climate movement. Unfortunately, victory through non-violent activism alone is not one of them. If only it were that easy to say that peaceful campaigners in the mid-nineteenth century brought about abolition and Reconstruction merely through skillful moral appeals and attention-grabbing but completely non-violent activism. But history tells us it wasn't so. Instead, it took abolitionists working together with the U.S. Army and Navy to free America's enslaved people. Ballots, yes. But also, sadly, bullets too.

Does that mean it will take the equivalent level of political activity on climate, perhaps in some way violent if not actually any kind of war or civil war, to abolish more than ten trillion dollars in oil wealth today? This is a serious question with important implications not only for those who see an analogy between slavery abolition and oil abolition but also for success in climate action in general. We'll return to this conundrum at the end of this chapter.

For now, let's talk about the very real possibility of wars, armed conflict, and civil unrest that unchecked climate change will bring worldwide and even perhaps inside the United States. In several spots across the planet, climate wars have already happened.

Without serious action on climate solutions, the world will certainly descend into more armed conflicts. At first, the effects on the American homeland may be largely indirect, as in the pressure to accept—or the decision to repel—refugees from drought-stricken Northern Mexico or from Caribbean island nations vulnerable to hurricanes. But as time goes on and climate impacts worsen both abroad and at home, levels of suffering only seen now in poor countries may be seen in more and more parts of the United States. The good news is that the nation's leading authority on national security threats is already on the case.

The Strangest Bedfellow for Climate Solutions

"We will have to pay for this one way or another," warns retired Marine Corps General Anthony Zinni. "We will pay to reduce greenhouse-gas emissions today, and we'll have to take an economic hit of some kind. Or we will pay the price later in military terms. And that will involve human lives. There will be a human toll. There is no way out of this that does not have real costs attached to it."[23]

The most experienced authority to talk about war and the threat of armed conflict is the military. In keeping with protocol under civilian command, the U.S. armed service leaders said little in public about climate change during the climate-denying Trump administration. But it's clear from statements made earlier under more science-friendly administrations, as well as muted comments under Trump's term of misrule, that all branches of the military recognize climate change as a serious factor in future conflict.

If you're a committed pacifist, as many abolitionists were before the Civil War and as most climate activists are today, it's eye opening to learn that the Pentagon has been studying climate impacts for years and making predictions on the implications for America's defense. If you want to reduce and then stop the increase of climate change while protecting America from the worst impacts of weird weather in the coming decades, it turns out you may have more in common with generals and admirals than you'd thought.

Independent defense analysts like Gwynne Dyer quoted at the top of this chapter contend that climate disruption alone will cause more deaths than all but the biggest wars. Serving commanders of the militaries of the U.S. and its allies have not gone that far, yet. Yet officers at the highest levels of command have recognized climate change as a "threat multiplier," a factor that will make armed conflicts from traditional causes more deadly and more difficult to deal with. "While climate change alone does not cause conflict, it may act as an accelerant of instability or conflict, placing a burden to respond on civilian institutions and militaries around the world," according to the NATO Quadrennial Review of 2010.[24]

The climate movement seems to be realizing that the military can be a powerful ally to build public support by promoting climate from an environmental issue, which is usually low priority for most voters, to one of national security, a high priority issue across the political spectrum. Take World War Zero, a climate action group formed in 2019 by former Secretary of State John Kerry, who was named top climate emissary by Joe Biden. Kerry's group is pushing for net-zero climate emissions worldwide by 2050. Unlike "net-zero" emissions pledges from oil companies that run counter to those companies' business model that requires them to keep producing fossil fuels, Kerry's effort is run by people whose business is national security.

Part of a broad coalition including political leaders from the U.S. and its allies, celebrities, and climate activists, the group's leadership draws heavily from the Pentagon. Coalition "enlistees" include former Secretaries of Defense Leon Panetta, Chuck Hagel, and Bill Cohen along with retired generals Stanley McChrystal and Paul Eaton. The latter wrote in the group's online magazine that "National Security and Combating the Climate Crisis Go Hand in Hand" and explained that "as someone who spent most of my adult life in uniform, commanding troops in war zones, I cannot be more clear: Renewable energy is security and security is renewable energy."[25]

The Pentagon stepping up as an ally for climate action, even in a quiet way, is one of the most promising developments in years. The U.S. military is the world's largest consumer of fossil fuels and the largest emitter of greenhouse gases. Senior officers know that the military has both an energy supply problem and a climate change problem. The Department of Defense can build out clean energy and increase energy efficiency on a huge scale—and they've already begun to do it.

It would be hard to find a more credible messenger on the dangers of climate change and the need for serious action on clean energy. The armed services remain some of the most popular and trusted institutions in American society, enjoying immense credibility with the voting public and commanding massive leverage in the economy.

Whatever you think of its war-fighting mission, that mission does require the Pentagon to accurately predict what the world's future will look like. Planning to adequately protect the U.S. from foreign foes doesn't allow for the luxury of sidestepping science for political reasons. Republicans in the White House like Trump and in Congress like perennial climate denier Senator James Inhofe of Oklahoma can spend years ignoring and denying inconvenient truths about the dangers of burning fossil fuels to keep their patrons in Big Oil happy. But the armed services are more reality based and don't have the luxury to play politics with potential threats to national security.

To the frustration of some independent defense analysts, as we discussed, the Pentagon is less focused on climate as a primary threat or a danger in itself than as a factor that can increase threats from traditional human adversaries. These include national governments in China, Russia, Iran, and North Korea, along with international terrorist networks like Al Qaeda. According to reports from the armed services over the last two decades, climate change will be a compounding factor that will make traditional threats more dangerous.

Refugees, Insurgents, and Great Powers in a Hotter World

In his book *All Hell Breaking Loose: The Pentagon's Perspective on Climate Change,* longtime defense analyst Michael Klare mines decades of reports by each of the U.S. armed services to show how the Pentagon has identified a variety of threats that are multiplied by climate change. Klare organizes them in an escalating "threat ladder" ranging from most to least likely.

Each step on this ladder is a threat that would divert military personnel and resources from their main mission of defending the American homeland from foreign adversaries. The higher you go on the ladder, the more dangerous the threat.

At the bottom of the ladder are found threats to American security that are most likely but least dangerous, like having to provide humanitarian disaster relief in foreign countries. As you climb the ladder the threats become less likely but more frightening. At the top you find threats that are extremely unlikely in the foreseeable future but would be most dangerous should they occur—ending with the "all hell breaking loose" scenario where multiple crises strike at the same time while the military's own bases are also damaged or made unusable by coastal storms, heat-driven power outages, or wildfires.

It's worth spending a little time to summarize each kind of climate-related challenge to understand how seriously the military takes climate change as a threat multiplier that could make traditional threats more dangerous.

1. Humanitarian Relief More Frequent and More Dangerous

Klare starts on the lowest rung of the threat ladder, how climate will impact missions for humanitarian disaster relief mounted by the United States military in foreign countries. In response to major hurricanes, typhoons, or other natural disasters, since at least World War II, the U.S. has undertaken relief operations as a gesture of friendship and compassion to friendly nations.

Aside from building goodwill, such missions offer a chance to show off the prowess of our armed services and keep personnel in practice. Disaster relief operations typically deploy American soldiers, sailors, pilots, and support staff to at the site of a natural disaster to deliver relief supplies and provide medical care and other emergency services until those can be taken over by local authorities.

Senior military officers worry that the number and scale of such operations could increase in the future, especially as climate change causes coastal storms to worsen. But that's not all. It's a cruel irony that increasing droughts and heatwaves that destroy crops will drive farmers from inland areas to seek employment in coastal cities—where they will add to the huge number of people already at risk from coastal storms. Making a bad situation worse, this cycle starts to drive itself.

While not every U.S. servicemember will enter combat, it's likely that most will be involved in a humanitarian relief operation at some point in their military career. While adding costs and pulling forces away from protecting the American homeland, humanitarian operations abroad also risk putting U.S. forces in harm's way. If one part of a local population feels that relief workers are giving more attention to other groups, then American military personnel could be pulled into civil unrest that could escalate into armed conflict.

And if natural disasters are bad enough, no amount of relief efforts can keep people from leaving a damaged place permanently, thus joining the explosion of refugees that represent one of the biggest threats to global peace and stability, a threat found on the next rung of the climate escalation ladder.

2. Friendly Nations Will Call on Us More Often for Help

Support for beleaguered foreign governments is the next level of threat about which the U.S. military worries that climate change could make more dangerous. Just as military planners expect to send more troops on missions to provide humanitarian relief abroad, so those planners also predict that friendly foreign governments will require more help to quell civil unrest among their population as severe droughts, widespread crop failures, and other climate impacts get worse and happen more often.

"Climate change can make already unstable situations worse, sometimes catastrophically so," according to Admiral David Titley. "Large-scale human suffering often accompanies these situations…U.S. military forces are frequently directed to these areas and our troops are placed at risk."

When foreign governments face armed insurgencies or are overthrown by violence, large numbers of citizens may flee to surrounding nations or further afield. If these refugees don't return home, they must compete for jobs and resources with often resentful local populations, creating tensions that can lead to more violence. Military leadership worries that if the local violence is bad enough and lasts long enough a country can descend into a protracted period of armed conflict depriving it of effective government. Such a failed state offers a political vacuum that can be filled by terrorist groups, as happened when Al Qaeda was able to set up operations in war-torn Afghanistan, allowing Osama Bin Laden to use the lawless nation as a base for the 9/11 attacks on the United States in September 2001.

3. Global Trade Disruptions Bring Oil Shocks and Food Shortages

The third and more deadly rung on the military's ladder of climate-influenced escalation involves disruptions to global supply chains. While one failed state may produce a limited wave of refugees or a localized opening for terrorists, as warming progresses, weather disasters can disrupt global trade in essentials like food and energy. Ultimately, to stabilize the climate, we must abolish oil. The process must be deliberate and serious. Yet to ensure a stable energy transition, phasing out fossil fuels must also take a decade or more to complete.

An oil shock or sudden cut-off of oil supplies as a result of a weather disaster, not to mention a similar crisis in delivering food supplies across international borders caused by climate impacts, could result in "widespread panic, flight, and instability," as Michael Klare explains. For the U.S. military, massive climate shocks could create a string of failed states "likely to trigger massive waves of human migration and help spread infectious diseases, producing disarray across the planet."[26]

The Arab Spring that started in 2010 and sparked protests in countries across the Middle East and North Africa including Tunisia, Libya, Tunisia, Egypt, and Yemen, may have been partially caused by climate impacts. Exceptional drought in China likely driven by global warming significantly cut grain harvests, which required the Chinese government to make up for domestic shortages through purchases on the global grain market. With deep pockets, the Chinese could afford to outbid traditional importers of food, including many countries in the Arab world, causing price spikes which in turn led to grain shortages in those poorer countries. This in turn led to bread riots and the overthrow of governments unable to feed their people.

As the climate gets hotter in the future, more harvests will come in short or fail altogether, putting even more pressure on global food supply chains. And that will certainly exacerbate political tensions, especially in dry countries with marginal agriculture that already import food. Food shortages can lead to civil unrest and topple governments, sending refugees out into neighboring nations causing political tensions that can explode into violence rippling across a region.

4. Nations Stake their Claims in the Melting Arctic

Discord among great powers such as the U.S., Russia, and China, especially in the thawing Arctic Ocean, is Klare's fourth level of escalation for climate as a threat multiplier.

For geostrategic analysts, the Arctic presents perhaps the world's greatest theater for potential conflict among great powers in the future. Klare thinks that military planners should be worried about melting ice caps and releases of methane trapped in permafrost, which could greatly accelerate climate heating. But senior officers are still more focused on traditional threats from foreign powers than from direct impacts of climate heating. However, military planners do recognize that as sea ice in the Arctic melts, a new ocean will open for the world's nations to use and to dispute.

Every year since 2006, the U.S. National Oceanic and Atmospheric Administration has put out the Arctic Report Card, updating the public on the effects of global heating on the part of the world that's changing more quickly than anywhere else.

The 2020 report had a scary new message, according to Henry Fountain, who has covered the Arctic in the *New York Times* for years. "It was not just that the Arctic is changing—that's been said umpteen times. It was that the region is shifting to a fundamentally different climate, that it is well on its way to becoming a place defined more by open ocean and rain and less by sea ice and snow. The Frozen North that we know is fading, and that will bring—and already is bringing—other changes far to the south."27

Already, all the nations bordering the Arctic are making plans to utilize newly ice-free sea lanes for navigation short cuts. Nations with a polar coastline also plan to exploit the region's resources in fossil fuels, minerals, and seafood. Polar states ranging from Norway and Denmark to Russia and the U.S. have all staked claims to exclusive use of Arctic waters far off their coasts, claims that potentially clash with each other. Making Arctic geopolitics even more crowded, in 2018, the Chinese government issued its first Arctic Policy, declaring itself a "near-Arctic state" and asserting its own interests in the melting northern ocean. All this sets the stage for diplomatic tensions and even armed clashes between the world's largest militaries in the future.

To make things worse, as climate change sends deadly heatwaves into already hot areas like the Middle East, nations where the heat regularly exceeds 100 degrees Fahrenheit for months at a time may become inhospitable to outdoor work of all kinds, including, ironically, oil drilling. If it's still legal for oil companies to expand into new areas, Arctic oil will become more valuable.

Meanwhile, heating oceans may send the world's leading fisheries into decline, increasing the value of food resources in the Arctic as well. Food fights and oil wars: War between major powers, hard to imagine today, is a real worry for U.S. military planners as Arctic ice shrivels.

5. Weather Disasters at Home

Near the top of the climate and conflict escalation ladder, a deadlier level of threat, are found domestic weather disasters. Even today, the world's richest countries suffer to a certain extent from hostile weather driven by climate change, whether it's wildfires in Australia, prolonged heat waves in Europe, or extreme storms along the U.S. Gulf Coast.

In the past, civil authorities in the U.S. asked the Department of Defense to help provide emergency health services and rescue people trapped by fires and floods, but the need for this was infrequent. Military leaders predict that the armed services will have to provide such responses more often in the future and that the military may be asked to respond to multiple disasters at the same time, as they had to do in the late summer of 2017 when Hurricanes Harvey, Irma, and Maria struck Texas, Florida, and Puerto Rico over the course of a few weeks.

Security experts also worry that, in coming decades, much of America's coastline will be permanently inundated by rising seas, displacing residents of major cities and destroying crucial infrastructure such as highways, rail lines, airports, refineries, power stations, and electrical transmission lines. Removing dangerous fossil fuel facilities from vulnerable areas is just another reason to get off oil.

Meantime, having to respond quickly to such disasters to offer relief and then afterwards, to help rebuild, will pull service people and military equipment away from defending the country against foreign threats, leaving the nation especially vulnerable to attack.

"If the environment deteriorates beyond some critical point, natural systems that are adapted to it will break down," writes former National Security Advisor Leon Fuerth. "This applies also to social organization. Beyond a certain level climate change becomes a profound challenge to the foundations of the global industrial civilization that is the mark of our species."

In his scenario for climate impacts on the U.S. homeland for 2040, Fuerth projects disaster across the country. In Western states, widespread drought and melting of snowpack in the Sierras and Rockies that feed rivers will lead to the collapse of California agriculture along with permanent water shortages in the cities of the Southwest. Rainfall will also decline over the high plains, but agriculture there can survive for a while by tapping the Ogalala Aquifer until heavier use speeds its depletion. Coastal populations in the heavily populated areas of the Southeast under constant attack from hurricanes will initially receive protection from federal relief projects for seawalls and runoff control.

Ultimately, the attempts will fail as brute engineering gives way to "strategic withdrawal" where gradual, planned relocation may degenerate into "sudden depopulation." Under all these stresses, as federal budgets are stretched, Washington will start to offload the costs of coping with ongoing relief to states, threatening the coherence of the nation itself, a nightmare scenario reminiscent of the tense era before the Civil War.[28]

6. War on Top of Disaster on Top of War

The top rung on Michael Klare's escalation ladder is "All hell breaking loose," an apocalyptic scenario where multiple disasters strike at the same time around the world. Weather disasters that hit` multiple countries at the same time would cause widespread instability and human suffering that will lead to food price spikes and shortages, massive flight of refugees, spreading of pandemics, rioting and civil unrest, collapse of governments, armed conflict between nations, and large numbers of people sick, injured, and dead.

In this worst-case scenario, even the wealthiest and strongest nations would not be spared. "The major powers, previously relatively immune to warming's harshest effects, could face catastrophic threats themselves and battle one another over access to critical resources," warns Klare.[29]

James Woolsey, former head of the Central Intelligence Agency, has been outspoken about climate change as a national security threat. If Greenland and Antarctic ice sheets melt, sea levels could rise two meters (six and a half feet), which scientists now say is possible by 2100.[30] In this scenario, even the wealthiest nations will hardly be able to look beyond their own salvation, writes Woolsey.

As nations struggle to deal with weather disasters, simultaneous climate impacts could merge into in one global "conflagration" of political and societal collapse: "Rage at government's inability to deal with the abrupt and unpredictable crises; religious fervor, perhaps even a rise in millennial end-of-days cults; hostility and violence towards migrants and minority groups, at a time of demographic change and increased global migration; and intra- and interstate conflict over resources, especially food and fresh water."[31]

What really strikes fear into the hearts of seasoned defense planners is that, while all hell is breaking loose across the world, back in the U.S., the military's own domestic bases could come under threat from storm surges, severe droughts, rising seas, and recurring wildfires. Compound that with the simultaneous collapse or destruction of civilian infrastructure like the power grid, roads, and off-base housing for military personnel, making it challenging for servicemembers to report for duty. "This threat to the Pentagon's installations at home, combined with an increased tempo of climate disasters abroad, conjures up the military's worst nightmare," writes Klare.[32]

A Greener, Meaner Fighting Machine

America's military leadership is clear that they see their role not as trying to prevent climate change or even to help make the nation more resilient in the face of weather disasters. The Pentagon's sole mission is to defend the nation from Russia, China, Iran, North Korea, and other foreign adversaries. At the same time, decades of Department of Defense reports show that senior officers fear that climate impacts will make it harder to accomplish their mission.

"Fighting against global warming is not, for most of them, a question of ideological preference or political engagement. Rather, it is a matter of ensuring that the armed forces will be spared the most extreme climate contingencies imaginable," to remain in fighting form to face America's foreign foes as Klare explains.[33]

The Pentagon understands that America's climate policy is largely a matter for civilian leaders to determine. But recognizing the threat multiplier of weather disasters both at home and overseas, the military is quietly acting on two fronts to reduce the damage while preparing for the worst. While helping prepare militaries of allied nations to better respond to climate disasters, all the armed services are trying to reduce their own contributions to greenhouse gas pollution by making their operations more energy efficient and switching to clean energy.

In 2011 the Department of Defense committed to reduce the energy intensity of its bases by 37.5 percent and to source 20 percent of its power from clean energy by 2020. With more than 200,000 buildings located on more than 4,000 different sites, this was an ambitious commitment made at a time that most other parts of American society were moving at a much slower pace to get off oil, gas, and coal. But there's still much to be done.

"Although the Department of Defense (DOD) has significantly reduced its fossil fuel consumption since the early 2000s, it remains the world's single largest consumer of oil—and as a result, one of the world's top greenhouse gas emitters," defense analyst Neta C. Crawford wrote in 2019.[34]

Meanwhile, each of the service branches has been acting to "unleash [themselves] from the tether of fuel," to quote General James Mattis. For instance, in 2016 the Navy launched its Great Green Fleet, a group of ships able to meet nearly half of their energy needs through renewables. The Marine Corps, which has committed to becoming "leaner, meaner, and greener" across all its operations, developed the Ground Renewable Expeditionary Energy Network System (GREENS), a portable solar and battery system large enough to power a platoon-size command center. And the Army is working to make its bases self-sufficient not only for energy but also for water supplies, designing 17 bases including the U.S. Military Academy at West Point as pilot locations for its drive to cut climate pollution.

While the military is trying to do its part to get off fossil fuels to bolster its own operations, their work will also serve as an example for the rest of the country and other nations as well.

The U.S. is the world's leading military power by far, making the Pentagon's example a powerful one for the militaries of the rest of the world. As to the civilian economy, because of the offshoring of manufacturing from North America and Europe to low-wage countries in Asia, especially China, that nation has become the world's biggest climate polluter. But China will have to pump out carbon dioxide and methane for years to come before it catches up with America's record as the globe's largest cumulative greenhouse gas polluter, a record built up over two centuries of industrial production. On the principle of "polluter pays," the United States has the biggest responsibility to clean up climate pollution. Since the U.S. boasts the globe's largest economy, and defense production is a huge part of our economy, then greening the armed forces will help the United States to show global leadership across industries from energy and transportation to housing and consumer products.

Famous stories of $800 toilet seats and other wasteful defense procurement aside, even a committed pacifist can't help but recognize that when the military wants to get something done, they have the money and manpower to complete the mission, to do it in a big way, and then to spread innovations to the private sector. From undershirts and feminine hygiene products to synthetic rubber and toy putty, to microwave ovens and GPS, to the internet and even solar panels, some of the most important inventions of the last century started in the military. Anyone who worries about climate chaos should support Americans in uniform in their efforts to protect themselves, which will help protect the rest of us in many ways.

Abolition Army, Army of the Green New Deal

Military planners make a frighteningly convincing case that climate emergencies will multiply threats to world peace and the stability of all nations as the world heats up. Though climate may not bring a new American civil war, continuing to burn fossil fuels is sure to make civil unrest and armed conflicts worse across the globe, bringing home to the United States first indirect consequences like an influx of refugees, and then inflicting direct damage that could engulf cities, states, and the federal government at home.

To reduce the suffering of cascading climate emergencies that could add up to World War III, the nations of the world, starting with the world's only superpower, the U.S., must take climate action on a much more accelerated schedule.

But to get that action, which will require us to abolish oil, the American people are up against an obstacle every bit as powerful and stubborn as the Slave Power of the Civil War era. Now we can return to the question of whether the future of oil abolition must follow the history of slavery abolition—violence and war. Given the wealth of Big Oil and their allies and how they use that wealth to control government and influence public opinion, will anything short of a new political conflict on the scale of the Civil War be enough to seriously transition away from fossil fuels and towards clean energy?

In nineteenth-century America, despairing of ending slavery by peaceful means, radical abolitionists did turn to violence. Enslaved leaders Gabriel Prosser, Denmark Vesey, and Nat Turner launched armed revolts that killed white and Black people alike, destroyed property, and ultimately, all ended with the execution of their leaders and the reimposition of control by slaveowners.

Before the Civil War, most white abolitionists were sympathetic to slave revolts, but remained committed to non-violence. But after decades of abolition activism that failed to end, reduce, or even stop the spread of slavery, pacifists came to appreciate that armed conflict may have offered the only remaining option to dethrone King Cotton and the Slave Power once and for all.

In October 1859, militant abolitionist John Brown led a party of 22 white and Black volunteers on a raid of the federal arsenal at Harpers Ferry in what was then Virginia (and is now in West Virginia). His crazy plan was to ignite a violent uprising of slaves across the South. After Brown's group was captured by U.S. Marines—ironically under the command of future Confederate Generals Robert E. Lee and J.E.B. Stuart—Brown was tried and hanged for treason to the Commonwealth of Virginia. Southern planters denounced Brown as a terrorist trying to incite slaves to slaughter their masters and their families while asleep in their beds. Indeed, most white people both South and North denounced Brown's use of violence.

But many slaves celebrated Brown as a would-be deliverer and abolitionists proved willing to put non-violence aside when violence was used in the service of abolition. Leading white abolitionist William Lloyd Garrison, "as an 'ultra' peace man," affirmed that he would have wished to "disarm" both Brown and his rebels along with slaveholders as well. Yet, after thirty years of waiting for slavery to end, Garrison felt compelled to praise the Harpers Ferry raiders: "Success to every slave insurrection at the South, and in every slave country."[35]

Frederick Douglass, the leading Black abolitionist, was a friend of Brown's and had met with the fiery zealot before his ill-fated invasion of the South. Douglass refused Brown's invitation to join the attack on Harpers Ferry, correctly predicting that Brown's little war would end in defeat and death. But after Brown was captured, authorities sought to arrest Douglass as an accomplice. Douglass fled his home in upstate New York for asylum in Canada, where he remained for a few months until President Buchanan's secretary of war—in another historical irony, Jefferson Davis, the future president of the Confederacy—announced that there would be no further arrests connected to Harpers Ferry.

Years after the war, Douglass eulogized Brown with the highest praise: "His zeal in the cause of my race was far greater than mine—it was as the burning sun to my taper light—mine was bounded by time, his stretched away to the boundless shores of eternity. I could live for the slave, but he could die for him."[36]

Garrison changed his attitude towards violence and ultimately learned to accept full scale war as necessary to end the evil of slavery. A year and a half after Brown's abortive raid on Harpers Ferry, when Confederates fired on Fort Sumter in April of 1861, Garrison predicted disaster for the Union and was ready to let the slave states go in peace. If Southern planters were allowed to secede, then the U.S. would finally be free of their influence over national politics.

By the end of the war four years later, Garrison changed his tune. After millions of slaves had been freed by the Emancipation Proclamation and an end to slavery in America was on its way as the Thirteenth Amendment made its way through the states towards ratification, the pacifist announced that "the American army was now the American antislavery society."[37]

On April 14, 1865, four years to the day after Fort Sumter surrendered to Confederates and the U.S. flag was hauled down, Lincoln arranged a ceremony at the bombed-out fort to mark the end of the war and Union victory. Among the honored guests who grasped the halyards to raise the old flag—which had been rescued from the surrender in 1861 and kept in a New York bank vault for the duration of the war—were Sumter's old commander Major General Robert Anderson; abolitionist preacher Henry Ward Beecher, whose sister Harriet Beecher Stowe wrote the antislavery blockbuster novel *Uncle Tom's Cabin*; and William Lloyd Garrison himself. The old pacifist had come full circle and recognized that it took more than moral suasion to defeat the Slave Power.

The American military today is hardly the army of clean energy. The Pentagon has overseen wars for oil in Afghanistan and Iraq. The U.S. Navy continues to patrol the Strait of Hormuz to keep oil from the Persian Gulf flowing to the world. Despite two decades of gains in energy efficiency and clean energy, the military is still the nation's largest consumer of fossil fuels. And while a few retired generals and admirals are willing to speak up about the climate threat to national security, serving flag officers are bound by protocol to refrain from advocating for public policy including climate solutions.

All this might make the military look more like the problem than the solution. Yet, through the power of its example to work progressively to abolish oil in its own operations, today's armed forces may help act to free America from the tyranny of fossil fuels just as U.S. soldiers and sailors in the 1860s morphed from forces tasked with putting down a rebellion into an army of liberation. At the end of the Civil War that finally brought freedom to the enslaved, victorious men both white and Black wearing blue uniforms were duly hailed as heroes by abolitionists, no matter how devoted those same abolitionists had been to non-violence for decades beforehand.

National Security Gets More Respect than the Environment

"Securitization" occurs when national leaders pluck some lucky political issue from the bowl of concerns that people and parties have argued about for years, promoting that one problem to a unique concern of national security.

In the United States, this happened when defense planners after World War II dubbed the competition for global influence between the Soviet Union and the West the "Cold War," or later in the 1960s when President Lyndon Johnson launched his "War on Poverty" to improve living conditions for residents of low-income neighborhoods and help the poor access economic opportunities long denied them.

Presenting a danger as "existential" will remove it from the normal sphere of politics and policies, writes defense analyst Anatol Lieven. "The threat is thereby placed in a special, exceptional category, backed by a national consensus and allowing the use of exceptional measures and the mobilization of national resources to meet it." That's why it will speed climate action if the U.S. military throws its weight, insofar as it is able, behind the Green New Deal. Lieven quotes from a report by the Army War College: "Army leadership must create a culture of environmental consciousness, stay ahead of societal demands for environmental stewardship and serve as a leader for the nation or it risks endangering the broad support it now enjoys. Cultural change is a senior leader responsibility."[38]

War is a crisis that speeds up politics. Can we create a similar crisis without war through peaceful but vigorous activism?

If climate remains a concern primarily of people on the political left, then climate solutions will never build the broad base of public support they need to go up against Big Oil. But with a consensus of the American public that climate is an existential threat to our own nation on par with a world war, then the movement for solutions can exert enough tension on political leaders to create a window of opportunity— a crisis—where actions previously unthinkable will now be possible. A revolution comparable to the Civil War without war, so to speak. That could lead to quick victories soon that few climate activists dare to imagine today.

The young white men of the North in the original Wide Awakes put on paramilitary uniforms with capes and marched in big cities to help Abraham Lincoln get elected in 1860. Then, when the Civil War broke out the following year, many of members of Wide Awakes clubs put on real army uniforms and marched out to battlefields to fight and die to preserve the Union. Along the way, they also helped bring freedom to enslaved people who had already begun to seize that freedom for themselves.

Today, young multicultural climate activists from across the nation with the Sunrise movement's Wide Awakes have put on their capes to campaign—nonviolently yet aggressively—that Americans and their political leaders wake up to the threat from climate change. It's a hopeful sign.

This time, the fight must be peaceful. Clear-headed people who care about climate should struggle to ensure that the United States never endures an armed conflict anything like the Civil War ever again. That means peacefully agitating for the massive transformation in the economy that it will take to get on the road to climate justice quickly enough to make a difference.

Peaceful activism wasn't enough to bring victory to abolitionists before the Civil War, but it could be enough to help the climate movement win today. Fortunately, over the past century, across the globe, peaceful activist movements have enjoyed more success than violent uprisings, according to Harvard researcher Erica Chenoweth. It doesn't require violence to make big changes in society today, but it does require a big commitment from enough citizens to force governments to act or to change. Studying thousands of non-violent civil resistance campaigns worldwide, Chenoweth found that in every case where a mass-resistance campaign attracted the "active and sustained participation" of at least 3.5 percent of a nation's population, the campaign has achieved its goal.[39]

In the United States, that would mean that it would take about 11.5 million people to actively support a campaign to abolish oil over a period of years to reach success. Chenoweth is careful to point out that past results don't guarantee future returns and there are exceptions to the 3.5 percent rule.

Another study we discussed earlier found that a vocal minority could achieve success in pushing social change to a tipping point once the issue got support from 25 percent of the public. Michael Mann, who has followed changes in public opinion on climate disruption over the years has concluded that the "issue public" for climate—those who prioritize the issue—has reached at least 36 percent of American citizens. That's a lot higher than the 25 percent threshold for support to blossom across society and comparable to the portion of Americans who supported marriage equality at the beginning of the Obama administration, just before the tipping point came which brought legislation expanding the definition of marriage.

Mann cites a group of climate experts who argue that climate advocates can push the public along by winning changes in taxes and regulations that are small in themselves but will "induce positive social tipping dynamics." Changes include ending fossil fuel subsidies and increasing incentives for solar power and energy efficiency; rebuilding cities with buildings and transportation that are carbon neutral; divesting from fossil fuel assets; and communicating the dangers and even moral bankruptcy of continuing to use dirty energy with better climate education and forcing industries of all sorts to disclose their greenhouse gas emissions.[40]

Whether they need to recruit 3.5 percent of the population as dedicated activists or merely reach 25 percent of the public as supporters, people who want to save the climate can take hope in these findings. We don't have to convince everybody. We just need to win a few strong allies and then engage a very doable population of Americans. Abolitionists were never more than a small minority in the nineteenth century. But they managed to win partial support from Free Soilers and other Americans who opposed the spread of slavery to new territories out West for their own reasons.

In any event, to move the federal government as well as state and local authorities to start abolishing oil now, the only practical path is non-violent campaigning along the lines of the civil rights movement.

As the pro-Trump insurrection of January 2021 showed, any violent alternative is unthinkable. And peaceful movements are doable. Despite stories of urban violence in the protests after the killing of George Floyd by a Minneapolis police officer in the summer of 2020, more than 93 percent of Black Lives Matter gatherings nationwide were nonviolent, with no injury to people and no damage to property.[41] Unfortunately, a few incidents of Molotov cocktails thrown, windows broken, and statues toppled gave Fox News enough footage to try to discredit the whole movement.

To keep the trust of a larger public, climate and racial justice activists alike must be vigilant about enforcing discipline among their members, rooting out or neutralizing outside provocateurs, and planning for multiple contingencies to reduce the risk that their activism will turn violent.

That America must take the lead to prevent the world from descending into climate wars goes without saying in an age when nine nations, including such unstable regimes as those in North Korea and Pakistan, can deploy nuclear weapons. With hurricanes, floods, wildfires, droughts, and heatwaves getting worse every year, soon, climate disruption may represent a more pressing threat then nuclear war. The United States and the world cannot afford anything like the doomsday scenario outlined by military analysts if runaway climate heating continues unabated: decades of hellish weather emergencies breaking out all over the place ultimately culminating in a worldwide conflagration, the Armageddon of global climate war.

To help avoid this dismal future, a later chapter offers a controversial idea from the abolitionist past that could help to peacefully abolish trillions of dollars of value in oil in the future without the need to fire a single shot.

For now, there's another factor that could make climate wars even more likely. Fossil fuel companies know as well as the U.S. military does that a degrading climate will make armed conflict more dangerous, and probably more frequent, in the future. And just like the military, Big Oil is planning to deal with a hotter world. Oil companies have stopped denying climate science and attacking scientists. But dirty energy companies haven't stopped trying to find ways to keep drilling, shipping, and selling fossil fuels.

We already saw how oil company BP has announced plans to cut its greenhouse gas pollution across its business and even to ramp down its production of oil in the future. Other oil companies have also announced similar plans to make their operations greener. Yet, these same companies also have plans to keep exploring for new reserves of oil and gas and to keep drilling more wells.

How can they do both at the same time? In the next chapter, we'll look at the oil companies' plans to continue doing business in dirty energy for decades to come. Then, we'll compare them to similar plans from Southern planters before the Civil War to also keep growing cotton, tobacco, sugar, and other commodity crops with slave labor.

5

Grow or Die

I want these countries [Cuba and Mexico] for the spread of slavery. I would spread the blessings of slavery, like the religion of our Divine Master, to the uttermost ends of the earth, and rebellious and wicked as the Yankees have been, I would even extend it to them.

—Albert Gallatin Brown, Senator from Mississippi, 1858[1]

It's the richest set of opportunities since Exxon and Mobil merged.

—Darren Woods, CEO, ExxonMobil Corporation, March 2018, on expanding production of shale oil in the Permian Basin, offshore oil in waters belonging to Guyana and Brazil, and liquefied natural gas in Mozambique and Papua New Guinea.[2]

When an industry is so evil or so dangerous that the rest of the world thinks it should be abolished, leaders of that industry have a choice to make. They can admit that the critics are right, apologize for the harm they've caused to the nations of the earth, and commit to develop a plan to phase out their operations. Or they can deny the criticism, disparage the critics, defy the will of the public, and keep on with business as usual.

The second option is what fossil fuel companies took when first confronted by climate science and activism. They decided not to put the safety of people and all life on Earth first and start an orderly shutdown of their business. Instead, they determined to put their profits first and fight back against their critics, and ultimately, against the world.

Starting in the late 1980s when climate change became a public issue, Big Oil chose to spread doubt about climate science, attack scientists and environmentalists, and dare anybody to stop them from doing business just as they pleased. By pumping big money into the pockets of politicians and PR flacks alike, dirty energy companies were able to control the federal government and confuse the public enough to allow them to continue to fight until well into the Obama administration or even later. Their mantra was simple: Climate change is a hoax. And for a long time, it worked. That was what climatologist Michael Mann in his book *The New Climate War* called the "old" climate war.

But recently, fossil fuel companies finally decided to change tactics, what Mann has termed the "new" climate war. Big oil companies stopped denying climate science, at least publicly, although they still secretly paid free market "think tanks" like the Heartland Institute and dinosaur politicians like Oklahoma Senator James Inhofe to keep spreading lies and confusion about climate science. But in their official statements of recent years, big oil companies like Exxon and BP started declaring that they accepted climate science wanted to be part of the solution to slow and stop climate heating.

At the end of 2020, ExxonMobil CEO Darren Woods said that "we respect and support society's ambition to achieve net-zero emissions by 2050." Exxon and other big oil companies issued plans promising to reduce their climate pollution.

Whether you believe what they say or not, Big Oil is now playing a different game. Instead of rejecting climate science, smearing the reputations of scientists, and ignoring demands from an overwhelming majority of the public to reform, oil companies are now promising to cooperate with climate solutions.

No surprise, there's a catch. Sure, they're willing to help with climate, but dirty energy companies still want to sell more dirty energy. So, even while he claimed that he respected society's commitment to cut climate pollution to zero by mid-century, Exxon CEO Woods also told investors that the company plans to expand its production both at home in the Permian Basin of Texas as well as in Guyana, Brazil, Mozambique, and Papua New Guinea.

These plans raise an obvious question: Which is it—will Exxon contract or will it expand? Will the company cut its climate pollution, or will it produce more oil and gas? Or will it claim that it can do both at the same time?

This is just one of the contradictions in the Oil Power's recent religious conversion on climate.

Though they have replaced the arrogant sneer of hard denial with the accommodating smile of soft denial, the oil executives and henchmen who Mann has dubbed the "inactivists" have no plan to dig up, sell, or burn less oil in the future. Instead, they have just come up with clever ways to wrap dirty energy in a better story. That story rests on the three legs of deflection, distraction, and delay, according to Mann.

Inspired by the gun lobby's motto that "Guns Don't Kill People, People Kill People," the climate inactivists have tried to deflect the blame from fossil fuel companies to consumers. We just sell the stuff because people want to buy it, they're essentially saying. You can't blame our product. You must blame the extravagance of consumer culture. If you want to cut climate pollution, then you need to change the whole structure of the economy and society, or at least convince consumers to use less energy.

The problem is that while saving energy and using less stuff is a good thing to do for many reasons, no amount of personal virtue will be enough to stop climate disruption, as we've seen. That will require government action to abolish oil. But that's exactly what inactivists are trying to avoid, or at least forestall, so that fossil fuel companies can continue to enjoy a few more decades of profits.

A deflection campaign also helps inactivists drive a wedge into the climate movement, leading activists of one stripe to direct their ire at activists with a slightly different approach instead of uniting against the real enemy, the fossil fuel companies.

Inactivists have gotten clever at exploiting preexisting divides between those who demand government action on the one hand those who preach personal conservation on the other. Internet bots of the Oil Power have stoked conflict on social media between vegans and omnivores, young people and Baby Boomers, suburban wilderness protectors and urban environmental justice fighters, and so on.

"Malice, hatred, fear, jealousy, rage, bigotry, all of the most base, reptilian brain impulses—corporate polluters and their allies have waged a campaign to tap into all of that, seeking to sow division within the climate movement and outrage on the part of their 'base'—the disaffected right," Mann explains. And all to the detriment of the climate movement by helping to ensure inaction for years to come.

Finally, the inactivists wage their new climate war with a couple of lies about science that they hope will discourage activism. The lies contradict each other. But no matter. It's not about changing anybody's mind. It's just about confusing people who care about climate enough to discourage action.

The first lie plays down climate change, claiming that the impacts on the weather of putting more CO_2, methane and other greenhouse gases into the atmosphere won't be that bad. Downplayers want us to think that it will be easy to adapt to the weather of the future. A few flooded basements in a few old houses can be pumped out. And if you can start growing orange trees in upstate New York or the Chicago suburbs, then there could be some upside too.

The second lie does the opposite, painting climate disruption as inevitable. Climate doomers claim that we've already pumped out so much greenhouse pollution that climate effects are locked in for centuries to come. Tipping points like methane releases from melting Arctic permafrost are now locked in. There's nothing anybody can do at this late date to stop runaway climate change. Doomers often come from the far left and claim to care deeply about the climate. But their predictions of inevitable doom wind up aiding and abetting inactivists by spreading exaggerated messages about climate impacts.

According to Mann, doomers portray "catastrophe as a *fait accompli*, either by overstating the damage to which we are already committed, by dismissing the possibility of mobilizing the action necessary to avert disaster, or by setting the standard so high (say, the very overthrow of market economics itself, that old chestnut) that any action seems doomed to failure. The enemy has been more than happy to amplify such notions."[3]

Mann's message is the opposite of what both downplayers and doomers would have us believe. The climate crisis is deadly serious but there's plenty of reason for hope if we keep fighting. In the past ten years his fellow climate scientists have determined that if the economy stops most of its climate pollution before too much damage is done, then the climate will soon stabilize. Mann wants people to be alarmed, but not so alarmed that they become paralyzed with fear or give up in despair. He is not such a stubborn fighter that he won't settle for less than going down with the ship. Mann is a realist who sees that the science offers a warning but also a way out. Alarmed but active is the proper response.

Swimming against the Abolition Tide

The Oil Power may want the same thing they always wanted, which is to sell as much dirty energy as they can with as few limits as they can get away with. But they've learned that today they need to put a happier face on their dangerous business than they had to in the past. At least when it comes to talk, Big Oil has gone from bold defiance to ingratiating diplomacy.

By contrast, the historical trajectory of Big Slavery's response to abolitionists was the exact opposite. At first, when the abolition movement started during the American Revolution, many slaveowners cooperated with abolitionists. But a few decades later, slaveowners did an about-face and decided to defy calls to free slaves. The story of why slaveowners switched their attitude to abolitionists from cooperative to confrontational in the nineteenth century can help us understand better why Big Oil made the opposite switch in recent years. Comparing the two switches of attitude can also predict what it will take to hold fossil fuel companies to their promises to help rather than resist the fight to save the climate.

It may be hard to believe today, but in the early days of the abolition movement, leading slaveowners and Southern political leaders were receptive to calls from abolitionists to end the transatlantic slave trade and even emancipate some or all slaves. Under British rule, all thirteen original colonies from Massachusetts and New York down to South Carolina and Georgia allowed slavery within their borders. "[H]ow is it that we hear the loudest yelps for liberty among the drivers of negroes?" quipped British literary icon Samuel Johnson in response to complaints about taxation without representation by restive American colonists in the 1770s.

By the time those colonists declared their independence from Britain, many founders of the new nation wanted to end slavery, if not immediately and all at once, then at least by phasing it out over time.

We've already seen quotes criticizing slavery from Founding Fathers, including even Southerners who owned hundreds of slaves like Thomas Jefferson. Talking against slavery while owning slaves wasn't just hypocrisy. Founders' actions genuinely challenged the institution of slavery, though in a moderate way that was acceptable for the politics of the late eighteenth century. They didn't institute or even argue for immediate abolition everywhere. But they did establish precedents to shrink and ultimately phase out slavery in the future. The biggest of these stakes in the ground for antislavery was writing that "all men are created equal" in the Declaration of Independence. But that founding document originally referenced slavery explicitly too. And not in support, but in criticism.

One of Jefferson's drafts of the Declaration blamed the British Crown for foisting the transatlantic slave trade on the American colonists in the first place, and then for vetoing several attempts later by legislatures in the colonial period to put a stop to traffic in kidnapped Africans. "He has waged cruel war against human nature itself," Jefferson complained of King George III, "violating its most sacred rights of life & liberty in the persons of a distant people who never offended him, captivating & carrying them into slavery in another hemisphere or to incur miserable death in their transportation thither." Jefferson went on to denounce the slave trade as "piratical warfare," "execrable commerce," and an "assemblage of horrors."

But unfortunately, this passage never made it into the final draft of the Declaration of Independence. As a compromise with delegates from the two largest slaveholding states, South Carolina and Georgia, the Second Continental Congress deleted this language before voting to approve the final text on July 4, 1776.[4]

During the Revolutionary War and its immediate aftermath, discussions about freedom and equality by patriot leaders emboldened early American abolitionists to speak up and start agitating for change for Black Americans. In April 1775, Benjamin Franklin helped start the first abolition group in the country, the Society for the Relief of Free Negroes Unlawfully Held in Bondage, in Philadelphia. Abolition societies would spread across the North and even into some Southern states over the following years.

Once the Declaration of Independence declared human equality to be a self-evident truth, Americans Black and white were inspired to question whether a free country could tolerate keeping so many of its people in chains. Starting with Massachusetts in 1780, where enslaved people brought a lawsuit for their freedom prompting the legislature to act against slavery as an institution in the state, each Northern state abolished slavery within its borders by 1804. It would take decades for all of them to complete phasing it out and for every enslaved person in those states to gain his or her freedom. At the time, some states in the Upper South also considered doing the same.

Abolitionists also pushed for action on the federal level, and they succeeded when the Constitutional Convention of 1787 included a commitment to end the transatlantic trade in slaves as soon as 20 years in the future. Congress duly outlawed the international slave trade at the earliest date allowed by the Constitution, which was January 1, 1808. Unfortunately, enforcement of the ban was criminally lax, allowing shippers whose greed outpaced their respect for the law to continue to illegally sail chained Africans to the Americas.[5]

But the commitment to end the slave trade was real. Throughout the nineteenth century many Southern slaveholders supported the ban on the foreign slave trade, especially those in states like Virginia and Maryland where lands were exhausted by decades of bad farming practices and agricultural production was languishing. For them, morality was combined with self-interest. In a trade that tragically split up enslaved families, slaveowners in the Upper South increasingly made money selling their surplus slaves in the domestic slave trade to booming cotton states like Alabama and Texas. States that exported slaves within the country didn't want competition from foreign importers of enslaved people.

Most leaders of slave states in the first fifty years after American independence conceded that holding people in bondage was inconsistent with modern egalitarian values and a modern economy. In an age of enlightenment when political thought was moving away from hierarchy towards equality, and the economy was diversifying away from growing crops on large plantations worked by slaves to trade and manufacturing, it seemed self-evident to nearly everybody that abolition was inevitable. Through the early nineteenth century, many Americans were willing to accept legislation that would gradually abolish slavery nationwide.

But this all changed when Eli Whitney invented the cotton gin in 1793 and this technology was adopted across the cotton states in the decades that followed. The gin speeded up the processing of cotton by a factor of ten and made profits from slave-driven agribusiness too good for planters and speculators to resist.

When you're in the commodities business, whether cotton or crude oil, you can make more money selling a lot of product cheap than a smaller amount of product more expensively. As white settlers pushed Native Americans off ancestral lands in new Southern states like Alabama, Mississippi, Louisiana, and Texas, opening up fresh land for slave-run agriculture, entrepreneurial cotton growers hoping to get rich quick changed their tune from acceptance and cooperation with abolitionists to denial and defiance. In this case, money shaped ideology.

In the face of huge growth in slave agriculture, spokesmen for big planters started to deny that there was anything wrong with slavery at all. As we saw earlier, Southern leaders after the cotton gin was invented even cast doubt on the Declaration of Independence, concluding that it was a mistake by the American founders to assert that all men were created equal. The reality, they claimed, was that the white race was destined to rule over Africans and their descendants. A new generation of Southern leaders acted as if they were unashamed of their peculiar institution and started to boldly assert that slavery was a "positive good" for both owner and owned. The triumph of money over morality was now complete.

Then, to protect their new money-making machine, big planters sought political cover from the federal government. This was a challenge because during the early nineteenth century, the population of free states gradually outpaced the population of slave states, potentially giving antislavery interests the ability to outvote the Slave Power in Congress and potentially abolish slavery nationwide. So, as new states entered the Union, Southern leaders insisted on maintaining an equality in the Senate between slave and free states.

For political leaders from slave states, slavery in America was no longer a regrettable leftover from a previous era that would eventually fade away and disappear in a new enlightened democratic political order. Instead, slavery had become the most beneficial way for white and Black people to live together, not just in the South, but anywhere else in the country or in the world.

In their bravado, planters converted a cause for shame into a point of pride. With so much more money at stake, slave state leaders adopted a new position of self-righteousness about their peculiar institution. Then they defied the rest of the nation to do anything to restrict slavery's operation or growth. Planters insisted on their right to own and use workers in chains at home in Southern states, to move their slaves to new territories out West, and to bring slaves with them wherever they traveled around the country, even in free states.

In the two decades before the Civil War, Southern leaders had ambitious plans to introduce slavery to new places. But slave state politicians usually kept expansion plans to themselves so as not to alarm Northerners. It's a lot like how oil companies today make plans to expand production but then try to obscure those plans with promises to become part of the solution for climate change.

Yet sometimes the truth will out. This happened with slavery expansion when, in a rare candid speech prior to the outbreak of the Mexican-American War in 1846, Virginia Governor Henry Wise urged Texans to "proclaim a crusade" to conquer the rich lands to the south, the states of Northern Mexico, and to "extend the bounds of slavery...and the result would be that before another quarter of a century the extension of slavery would not stop short of the Western [Pacific] Ocean."[6]

Other Southern leaders cheered on mercenaries who launched freelance attacks intended to set up slave-friendly republics in parts of Mexico, Central America, and the Caribbean.

After American victory in the Mexican War in 1848 brought California and other areas of the southwest into the United States, Southern military adventurers known as filibusters (a term later adopted for the parliamentary procedure to delay a vote in the U.S. Senate) led freelance invasions attempting to "liberate" Spanish Cuba, Nicaragua, and parts of Mexico.

Their hope was to seize and subjugate these foreign lands where slavery had already been abolished. Then filibusters would set up governments led by Americans that would reinstitute slavery, thus opening even more new lands for white planters to produce lucrative commodity crops like cotton and sugar on the backs of enslaved men and women.

The most famous filibuster was William Walker, a child prodigy from a prominent family of slaveholders in Tennessee who received a law degree at age 14 and a medical degree at age 19. In his early twenties, Walker followed the gold rush to San Francisco, where he edited a newspaper and fought three duels. But his heart remained in the South and with its economic model of selling as much cheap cotton and sugar as possible, produced with the cheapest labor available.

Walker conceived an ambitious plan to conquer vast regions of Central America to add new slave states to the Union. That would increase both the size and the political power of his native Dixie. In October 1853, Walker led 45 men, many of them supporters of slavery from Kentucky and Tennessee, into Mexico. Walker's little army captured La Paz, the capital of the sparsely populated Mexican state of Baja California. Then, Walker declared an independent republic governed by the laws of the U.S. state of Louisiana, making slavery legal where it had been prohibited previously when Mexico abolished slavery in 1829. After three months, Mexican troops forced Walker to retreat north across the American border, where he was put on trial for conducting an illegal war that violated U.S. law. But filibustering was popular among Americans in Southern and Western states at the time. Nobody was surprised when a California jury took only eight minutes to find Walker innocent.

The brash Tennessean was beaten but not discouraged. A year and a half later, hearing that fighting had broken out between two elite factions in Nicaragua, Walker sailed from San Francisco in May 1855 with 60 men to try to take control of the Central American nation. With better luck this time, Walker toppled the government and declared himself president. His filibuster administration repealed laws forbidding slavery and ruled Nicaragua for a year and a half until Walker was defeated by a coalition of Central American governments in May 1857.

Walker still had one more fight in him, and he returned to the region yet again a couple years later. This time, Walker's luck had finally run out. Captured by the British Royal Navy and turned over to the authorities in Honduras, the mercenary leader was executed by firing squad in September 1860 at age 36. Walker had lived a short but colorful life dedicated to the despicable cause of making history go backward through freelance imperialism that would return freed people in Latin America to chains.[7]

Wildcatters: The Filibusters of the Oil Industry

Respectable Southern leaders in the White House or with high positions in the federal government in the decades before the Civil War usually distanced themselves from Walker and other mercenaries who waged illegal private wars on nations at peace with the United States. But Walker became a hero to many white Southerners who wanted to see the field open to the cheapest form of mass labor, slavery, spread beyond the American South.

To follow our analogy from slavery to dirty energy, in its early days, the young oil industry was also driven by a kind of fortune-hunting pioneer reminiscent of a filibuster. "Wildcatters" were freelance prospectors who led gangs of eager men to spots at home and abroad in search of the next big strike of crude oil. Like the filibusters of earlier decades, many of the wildcatters of the late nineteenth and early twentieth centuries were colorful adventurers unafraid to venture into the most forbidding desert and jungle in search of quick riches. Willing to take big risks, wildcatters used drills and oil rigs to seek oil in unproven fields to expand the domain of American oil production just as filibusters brought rifles and cannons to expand the domain of Southern planters to foreign lands.

The lives of wildcatters offered iconic American stories of rags to riches, with personalities every bit as colorful as that of filibuster William Walker. The first to successfully drill for oil in the U.S. using modern methods of sending a drill through a metal pipe was "Colonel" Edwin Drake. He struck crude in Titusville, Pennsylvania on the of the Civil War in 1859. But far more colorful were the wildcatters who appeared as word of Drake's discovery spread and set up boomtowns such as Petrolia and Pithole.

Among the young men who flocked to Pennsylvania oil fields during the Civil War was a struggling actor and Confederate sympathizer from Maryland named John Wilkes Booth. He joined a few friends in the arts to form the aptly named Dramatic Oil Company and acquire some promising land. Booth spent several weeks trying to strike oil on his property all the while entertaining the locals with an equally dramatic rendition of the Sermon on the Mount. By the end of 1864, Booth's oil well was dry. Disappointed yet again, Booth departed for Washington, DC, where he would begin organizing a conspiracy targeting the highest officials of the federal government. On the evening of April 14, 1865, Booth fatally shot President Lincoln at Ford's Theater.[8]

As the Pennsylvania oil patch started to come under the control of John D. Rockefeller and his Standard Oil trust, small independent operators were squeezed out of America's first oil market. Seeking opportunity free of big corporate control, wildcatters less sinister than Booth but just as eager to get rich quick set their sights on Western states that were still virgin territory for the industry and offered opportunities to discover new oil reserves.

Born during the Civil War and raised in the rough Texas lumber town of Beaumont, Patillo Higgins was known as "Bud," a prankster who spent his time drinking, gambling, and getting in fistfights. This bad hombre went on to harass a Black church and killed the sheriff who confronted him. Higgins was also shot in the gunfight. The injury wasn't fatal, but Higgins's left arm was so mangled that it had to be amputated. An all-white jury ruled that Higgins was just defending himself. After this exoneration, Higgins changed his ways, and at age 22, he accepted Jesus at a fire-and-brimstone revival at Beaumont's opera house. The one-armed Higgins then tried his hand at different jobs, including a tour in the oil fields of Pennsylvania, where he developed an intuitive ability to "read the land" and guess where oil might lie underneath.

Returning to his native Beaumont, Higgins partnered with "Captain" Anthony Lucas, an engineer who had served in the navy of Austria-Hungary. Together, the duo made the biggest oil strike ever up to that time at Spindletop near Beaumont in 1901, which began the Texas oil rush and made oil plentiful enough not merely to be used to lubricate machine parts but to be burned for fuel. Gulf Oil and Texaco were both formed to develop oil at Spindletop.[9]

An even bigger oil strike came in 1930, when 70-year-old Columbus "Dad" Joiner discovered the East Texas Oilfield, the largest petroleum deposit identified up to that time. Born and raised in Alabama, Joiner learned to write by copying text from the Book of Genesis. An appealing scoundrel, Joiner later used his writing skills to offer his condolences to recent widows that he discovered through newspaper obituaries. He contacted these bereaved wives less to soothe their grief than to help relieve them of any inheritance from their late husbands.

While more established oilmen like John D. Rockefeller looked down on wildcatters, these oilfield adventurers became heroes to would-be entrepreneurs across America eager to seek their fortunes in the booming business of oil, just as filibusters had become heroes for men eager to earn glory and riches in foreign conquest.[10]

Oil did make a few men rich quickly, while leaving many others even poorer than they were before. But even from its early days, oil was a dirty and dangerous business, fouling lakes, rivers, and creeks; denuding hillsides of trees which led to erosion and landslides; and starting well fires, the industry's worst plague. Whether sparked by thunderstorms or carelessly discarded cigars, the early years of oil in Pennsylvania witnessed numerous fires both in the oilfields and at refineries.

"Flames...leaped thirty meters high," described a witness to a well fire that burned out of control, "and for two weeks the fire raged, consuming ten acres. The fire could be seen for kilometers, and vegetation was scorched in all directions." And when a refinery exploded in Philadelphia in 1865, killing dozens of people, reporters described a hellish scene. "The blazing oil that escaped from the burning barrels poured over into Ninth Street…filling the entire street with a lake of fire…men, women, and children were literally roasted alive in the streets."[11]

Scenes of suffering straight out of Dante's *Inferno* were repeated in Pennsylvania's Petrolia region and then, as the oil industry grew, in the oilfields of Texas. But dangerous conditions didn't stop money-hungry adventurers from trying to make their fortune by expanding the empire of oil even further.

With wildcatters paving the way, the American oil industry would later strike out beyond the early oil patch states of Pennsylvania and Texas to some of the same foreign lands that the old filibusters coveted in Mexico, Central America, and the Caribbean. But wildcatters lacked the resources to find and develop the mammoth oil resources around the world that became synonymous with the global market in crude oil. It would take major American and European corporations like Exxon, Shell, and BP to turn Venezuela, Nigeria, Russia, Iran, and Saudi Arabia into the oil exporting giants they became in the twentieth century.

Likewise, back before the Civil War, while filibusters exhibited daring and ambition, it would take more established powers to support the international slave plantation economy in the largest foreign lands where it flourished in the Western Hemisphere. It would take the U.S. government itself.

Making the World Safe for Slavery

Prior to the election of Abraham Lincoln in 1860, Southerners exerted control over the federal government disproportionate to the size of the South's free population. This was partially due to the three-fifths clause in the Constitution that gave each slave state extra representation in Congress and in the Electoral College.

"Since the slavery agitation," remarked an antislavery congressman from Iowa at the end of the Civil War in 1865, slaveholders "have had the Secretaryship of State for two thirds of the time; and…for four fifths of the time have the Secretary of War and the Secretary of the Navy been from the South." [12]

Control in Washington, and particularly over what one historian calls "the outward state"—the diplomatic corps and military—allowed Southern leaders in Washington to convert foreign policy for the whole

United States into a tool for protecting and expanding the economic driver of one region, the South. And while Southerners insisted on a strict respect for states' rights at home to allow slavery, when it came to foreign policy, Southerners had no problem with big government bending other nations to its directives, if that would protect and promote slavery abroad. In fact, they insisted on it.

In the next chapter, we'll see how the British Empire peacefully emancipated its slaves in the 1830s. For now, we only need to understand how abolition in Britain's colonies in the Caribbean galvanized U.S. foreign policy before the Civil War by providing the Southerners who had an outsized influence in Washington an excuse to mount an aggressive proslavery foreign policy.

In the decades after the U.S. won its independence from King George III, Americans feared the power of British imperialism to threaten or even reverse American independence. Presidents from George Washington onward viewed the British Crown's remaining colonies in the Western Hemisphere from Canada to the Caribbean as potential launching pads for an invasion of American territory, fears which were realized when the two nations went to war again in 1812. A British force wound up capturing Washington, DC in 1814 and burning federal buildings including the Capitol and the White House.

Though British colonies in the Caribbean still allowed slavery, just as they had done previously during the American Revolution, in the War of 1812 British forces offered freedom to any enslaved American who would flee his master and join His Majesty's forces. After the war, the idea that Britain could foment a slave rebellion in the United States would continue to send chills down the spines of white people across the South, whether they owned slaves or not.

In the Revolution and the War of 1812, British forces offering freedom to slaves of American owners wasn't about abolition but just a military tactic common during wars at the time to weaken the enemy. But only two decades later, after abolishing slavery in its Caribbean colonies, Britain would start to exert its military and diplomatic power for abolition.

As we'll see in the next chapter, in response to an energetic campaign by abolitionists and a massive slave revolt in Jamaica, Parliament freed slaves on that island as well as in the Bahamas, Guyana, and other British colonies in the Caribbean in 1833. With the greatest military power in the world at the time now on the side of abolition, American slaveowners started to worry that British imperialism would become a force to free slaves throughout the hemisphere, including in the American Southern states. Such fears turned out to be overblown and Southern leaders overestimated the British government's commitment to spreading emancipation across the Americas.

Nonetheless, after the British Empire abolished slavery, when the issue came up in various diplomatic issues with the U.S. and other slaveholding countries in the Western Hemisphere, British colonial officials, consular agents, and Royal Navy commanders tended to side less and less with slaveowners and more and more with the enslaved.[13]

In the first half of the nineteenth century, U.S. foreign policy was largely directed by Southerners and their proslavery allies from the North. John C. Calhoun, the "cast-iron" advocate for slave planters from South Carolina, served at different points in his long career as secretary of war, secretary of state, and vice president.

During this period, slaveholders and compliant Northerners like New Hampshire's Franklin Pierce and Pennsylvania's James Buchanan occupied the White House. These proslavery presidents oversaw one compromise after another during the years before the Civil War in a doomed attempt to prevent sectional conflict over slavery. Compromises like the Fugitive Slave Act of 1850 that attempted to mollify tetchy Southern planters by making slavery more secure where it already existed and easing the potential spread of slavery to new places angered Northerners and only put off the day of reckoning between North and South.

Under proslavery management, the U.S. State Department sought to counter alleged British influence in independent nations and European colonies across the Americas. This concern especially applied to Texas after it broke away from Mexico in 1836 and before the Lone Star Republic joined the United States during the Mexican War in 1845.

During the decade that Texas existed as an independent slaveholding republic, Calhoun and other Southern leaders in Washington worried that British agents would intrigue to abolish slavery there. The Republic of Texas was offered a loan by the British that would have allowed the independent country to carry on, but that, since the loan required the abolition of slavery, Texas leaders decided to give up their independence and become a state of the Union in order to maintain slavery.[14]

United States annexation of Texas, which was guaranteed by American victory in the Mexican War, put Southern worries about British interference to rest, as London surprised leaders in Washington by granting official recognition to American rule over Texas and other territories taken from Mexico. Slaveocrats breathed a sigh of relief and looked forward to great things in the future. One novelist from South Carolina whose political views were popular with Southern leaders exulted in the march of U.S. armies into Mexico City and noted that "our Mexican conquests" would secure "the perpetuation of slavery for the next thousand years."[15]

With worries about Britain as a force for abolition abated after the Mexican War, U.S. foreign policy, still under the helm of Southern slaveowners, found a new threat to the future of slave-powered agriculture: foreign nations and colonial powers that might buckle to abolitionist pressure and decide to emancipate their own slaves. American diplomacy sought to protect slavery in the two foreign places where bound labor was still allowed in the Caribbean and South America—the Spanish colony of Cuba and the independent nation of Brazil. If either place were to abolish slavery, Southern leaders worried, it would only embolden abolitionists to demand the end of slavery in the American South.

When fears were stoked in the early 1850s that the Spanish government was considering abolishing slavery in Cuba, Southern leaders who loudly defended states' rights at home called for the opposite abroad, urging an American invasion of the island with the express purpose of stopping abolition there. Convinced that "if slave institutions perish [in Cuba] they will perish here," a former governor of Mississippi even raised a Walker-style private army of mercenaries to stop the Spanish from emancipating Cuban slaves. A South Carolina newspaper called for more official action to stop the "Africanization of Cuba": "If our interests are imperiled, let the Army and Navy be summoned to their duty."

Apparently, the small government those Southern leaders constantly preached was only good at home, an excuse for stopping the federal government from investing in such services as schools or help for farmers. In foreign affairs, at least when it came to protecting slavery abroad, only the biggest kind of government, military intervention, would suffice. Accordingly, in the years before the Civil War, Southerners in the federal government were the biggest champions of defense spending, calling for more and better troops, guns, and ships.

Just a few years before he would be elected by his fellow secessionists as president of the Confederacy, Mississippi's Jefferson Davis worked diligently to boost the fighting power of the United States during his term as secretary of war under President Franklin Pierce from 1853-1857. Having no idea that Southern states would soon leave the Union, and assuming that the federal government would continue to use its military to support slavery abroad for the foreseeable future, Davis raised pay scales for the first time in 25 years and increased the size of the regular army from 11,000 to 15,000, while also purchasing more of the type of rifles that had proved valuable in the Mexican War.

At the same time, New England abolitionists called for cuts in federal military budgets. Ironically, those soldiers and weapons that Davis and other Southerners acquired for the U.S. military while serving in Washington would soon be turned against those same Southerners only a few years later in the Civil War.[16]

As the abolition movement gained momentum in the 1840s and 1850s, champions of slave-powered plantation agriculture grumbled about alleged bad treatment by Northern leaders and persecution by antislavery activists. A few Southern hotheads even threatened secession, but more measured leaders like Davis counseled patience and compromise. Senior slave-state politicians in Washington still felt that the South could better protect and expand slavery inside the United States with all the resources of American economic and military power to draw upon than the South could do on its own.

But that changed when Abraham Lincoln won the hotly contested presidential election of November 1860. Governors and members of legislatures of Southern states from South Carolina to Texas immediately began talking openly about secession. In their mind, the antislavery Illinois lawyer was not their president. All the 180 electoral votes that gave Lincoln victory came from free states. And ten Southern states contributed not even one of the 1.8 million popular votes cast nationwide for the Republican who promised to contain the spread of slavery.

White Southerners saw Lincoln's election as a repudiation of the contract between North and South that both allowed and encouraged slavery for decades. "The Union of the present day is not the Union of our fathers," opined a San Antonio newspaper a couple weeks after the election. "It has been utterly perverted from its original design, and it has become an engine of oppression, wrong and tyranny, which those fathers never contemplated."

Within a month after Lincoln's election, every Southern state except Texas, restrained by pro-Union governor Sam Houston, had either planned to call or had already called a convention to consider secession.[17] "Fire-eating" radical proslavery Southern politicians urged secession as the only way to protect Southern rights. And those rights included not just the ability to keep slaves in the South, but also the freedom to take their slaves anywhere in the country, especially out West.

As former South Carolina Senator Robert Rhett promised in a speech intended to whip the citizens of Charleston into a secessionist froth right after Lincoln's election, without Northerners and abolitionists to stand in the way, an independent Southern nation could dedicate itself to extending slavery throughout the Western Hemisphere.

"We will expand, as our growth and civilization shall demand— over Mexico—over the isles of the sea—over the far-off Southern tropics—until we shall establish a great Confederation of Republics— the greatest, freest and most useful the world has ever seen." By "freest" Rhett meant that his new William Walker-style republics would be free to own slaves if their American conquerors wanted to. Black people, of course, would not enjoy any of Rhett's special kind of freedom.[18]

Lincoln tried to reassure Southern leaders that he was only elected to stop slavery from spreading to new Western territories while promising not to touch slavery in the states where it already existed. But this reassurance did not satisfy leading slaveowners. Without the chance to grow and expand the cotton and sugar economy to new territories out West or even south into Central America and the Caribbean, the value of planters' slaves would start to decline. That would seriously cut into slaveowners' wealth. And within the U.S., if new states were added that excluded slavery, it would tip the balance of power in the Senate towards free states, potentially allowing Congress to pass abolition.

"Expand or Die" should have been big slaveowners' motto. So, after Lincoln's election, cotton-state leaders were not going to wait around and see how things turned out under the new antislavery administration that would soon be coming to the White House. Instead, Southern state governments decided to get out of the Union while the getting was good. Having lost with the ballot, they were prepared to resort to the bullet.

Grabbing for Dollars on the Way out the Door

As one Southern state after another announced that it was leaving the United States in the months after Lincoln's election, local proslavery militias tried to seize federal property, especially arsenals, military bases, and coastal forts.

During the early months of 1861 leading up to the outbreak of hostilities in April, dozens of senior officers from Southern states resigned their commissions in the U.S. Army or Navy to offer their services to the Confederacy. In an especially brazen move, one of these fleeing officers, Pierre Gustave Toutant Beauregard, had the nerve to demand that the federal government cover his expenses to travel back to his home in New Orleans. Beauregard would soon command the Confederate artillery in Charleston that fired on Fort Sumter in April 1861 and unleashed four years of bloody civil war.

Fast forward to 2020, and another presidential transition led to another bold move by brazen leaders defeated at the polls just as they were on their way out the door. A couple weeks after Donald Trump lost his bid for a second term in the White House to Joe Biden, the lame-duck president announced that the federal government would auction off the rights to drill for oil and gas in the Arctic National Wildlife Refuge.

Established in 1960, the pristine wilderness area located on Alaska's North Slope is as big as the state of Delaware, making it the largest national wildlife refuge in the United States. Home to animal species including polar bears, Black and brown bears, caribou, moose, eagles, and many species of migratory birds, the area also contains up to 11.8 billion barrels worth of oil and gas. After 40 years of disputes between oil drillers and environmentalists, with Native Alaskans on both sides, in 2017 the Republican-controlled Congress passed legislation opening the Alaskan coastal plain to oil and gas development.

Citing this law, in November 2020, just after Donald Trump lost the presidential election to Joe Biden, the Trump administration started a rushed process to get drilling leases approved and sold before it was due to leave office in January of the following year.

"The Trump administration is hell-bent on selling off the Arctic Refuge on its way out the door, rules and laws be damned. But they have made a complete mess of the leasing process," said one opponent of drilling. Another agreed: "You're one mile from the finish line and you decide to take a shortcut is what this screams to me and you hope you don't get caught."

Kara Moriarity, head of the Alaska Oil and Gas Association, recognized that oil companies would have less time than usual to prepare their bids, but she didn't see anything rushed about the lame-duck Trump administration process to open the protected area. "From an industry standpoint, having a lease sale is the first step of getting access to responsibly developing resources that are needed to meet global demand," she said.[19]

Voters who cast ballots for Joe Biden said that climate was one of their top issues and that they prioritized transitioning the economy to clean energy. In this, they agreed with an overwhelming majority of Americans, both Republicans and Democrats, those who vote and those who don't. It's not surprising that Trump didn't cater to voters who supported his opponent. But to see the fossil fuel companies act so brazenly against the expressed will of voters who want action to stop climate disruption contradicts claims that Big Oil has made recently that they want to be part of the solution on climate change. It's just another example of the dishonesty of an industry that has said one thing and then done another for more than a century.

In addition, knowing that Biden was opposed to drilling in the Arctic National Wildlife Refuge and would certainly stop the process when he took office, the industry was apparently playing a deep game. It didn't help them win when, just after Trump invited oil companies to submit their bids to drill, Bank of America announced that it would join every major U.S. bank in refusing to provide financing for any drilling in the wildlife refuge. For climate activists, it was a hopeful sign of success for their new strategy of applying pressure on Wall Street to "stop the money pipeline" of financing for new fossil fuel projects.[20]

Predictably, after just a few months in office, the Biden Administration put the leases on hold, pending an investigation that could cancel them altogether. But the industry and its allies in government weren't worried.

"Since the Carter administration, whether ANWR can be leased is determined by which party is in the White House," said Marcella Burke, an energy policy lawyer who served in the Interior Department during the Trump administration. "Developers in ANWR assume there will be a policy shift between Democrat and Republican administrations. But it's not permanent, assuming there will someday be another party in the White House." Since presidential action can be undone, environmental groups called on the administration and Congress to permanently ban drilling in Arctic areas.[21]

Big Oil is nothing if not resilient. The industry has learned how to outlast presidential action against them as it learned to work with a Republican president that the industry didn't chose, Donald Trump. Major oil companies didn't give much support to Trump's presidential campaign in 2016, because, like so many smart people, fossil fuel CEOs assumed that Hillary Clinton would win. But once Trump was elected, Big Oil quickly allied with him to get as much as they could out of Washington.

Chevron, Exxon, and BP each donated at least $500,000 to Trump's inauguration in 2017, and of course they expected payback. And Trump did his best to oblige. When he entered the White House, Trump signed one measure after another to roll back environmental protections and help the industry, whether with new opportunities to drill or with coronavirus relief funding for fossil fuel operations. It seems that the oil industry got good value for their money: Over four years, Trump rolled back more than 125 environmental regulations, with many rollbacks occurring just under the wire in the last couple months of his lame-duck administration.

When it came to protecting clean air and water, Trump was the worst president in decades, including recent Republicans. But if you think that Trump came up with the idea on his own to cancel dozens of arcane regulations on such subjects as greenhouse gases, coal ash waste, water pollution, mercury levels, and smog, then you're giving him too much credit. Trump didn't know or care much about policy issues except immigration, trade, and defense. On energy, as with so many other issues, Trump outsourced public policy to the Republican donor class.

At the top of the list of generous and loyal Republican donors was Big Oil. Though they didn't pick Trump as the winner in the 2016 campaign, dirty energy producers stayed loyal to Republicans under Trump's regime. "Led by the oil and gas industry, [the fossil fuel] sector regularly pumps the vast majority of its campaign contributions into Republican coffers. Even as other traditionally GOP-inclined industries have shifted somewhat to the left, this sector has remained rock-solid red."[22]

Charles Koch, the richest oil billionaire in the United States, claimed that he didn't support Trump for president. But after Trump's surprise victory in 2016, Koch wasn't afraid to ask for a lot of help for the oil industry, even before Trump was inaugurated. In early 2017, a Koch front group called Freedom Partners published a detailed agenda for the coming administration, "Roadmap to Repeal: Removing Regulatory Barriers to Opportunity," listing the laws and regulations it expected to be repealed in the first 100 days of the Trump White House. And when Koch's people talked, Trump's people listened.

"The Trump administration dutifully marched to the beat," according to campaign finance watchdogs Pam and Russ Martens. "Repeal the Paris Climate Accord–done. Tax cuts for corporations and the wealthy–done. Gutting federal regulations and the Environmental Protection Agency–done."[23]

On other wishes of oil, gas, and coal companies, Trump didn't waste any time paying back fossil fuel donors. First, in December of 2016, only two months after the election, President-elect Trump nominated former Exxon CEO Rex Tillerson for the most prestigious post in his cabinet, secretary of state. Then, shortly after taking office, in March 2017, clearly drawing from the Freedom Partners list of environmental regulations to repeal, Trump published his Presidential Executive Order on Promoting Energy Independence and Economic Growth, targeting any federal regulations "that potentially burden the development or use of domestically produced energy resources, with particular attention to oil, natural gas, coal, and nuclear energy resources."[24]

The goal was not just to achieve "energy independence," as under the previous oil-friendly Republican administration of George W. Bush. Instead, Trump and his cabinet touted a more macho version they dubbed "energy dominance."

As Trump's first secretary of the interior, Ryan Zinke, explained, "there is a difference in energy independence, and there is a difference in energy dominance. We're in a position to be dominant. And if we, as a country, want to have national security, and an economy that we all desperately need, then dominance is what America needs." Translated, that meant to promote as much oil, gas, and coal development as possible.[25]

Trump was so aggressive about helping fossil fuels that it was like the Bush Jr. administration on steroids. Kathleen Sgamma, president of the oil and gas trade group Western Energy Alliance, expected Trump to be good for fossil fuels. But she was pleasantly surprised at just how much Trump wound up catering to dirty energy interests. "Not in our wildest dreams, never did we expect to get everything."

"Everything" in this case meant putting fossil fuel lobbyists at the head of the line for administration jobs and making it top priority to review and rollback restrictions on drilling and mining.[26]

Big Oil Will Be Forever Trumpers

Just as big cotton growers in the mid-nineteenth century American South firmly tied their fate to the Confederacy, so big oil, gas, and coal companies went all in for Trump. Let's hope that Trump will turn out to have been the last fossil-fuel president.

Americans should never forget the hell that oil companies and the PR men paid to spread lies about climate science by oil companies were willing to put the country through by enabling Trump. It all ended with Trump provoking the insurrection of January 6, 2021, the first attack on the U.S. Capitol since the War of 1812. And all just so that they could protect their profits.

"Climate deniers backed violence and spread pro-insurrection messages before, during, and after January 6," according to *DeSmog*. The evening after the insurrection, former coal mining executive Don Blankenship, who spent a year in jail for violating mine-safety standards while CEO of Massey Energy, tweeted his support for the insurrectionists. "Members of the media and the government are all saying what we saw today doesn't work—but that is only because they don't want it to work. What we saw today is what freed Americans from King George and England."

Tom Tanton, a consultant for the American Petroleum Institute and advisor to the Heartland Institute, one of the front groups discreetly paid by oil companies to deny climate science, falsely claimed on Facebook that the rioters were not Trump supporters, but infiltrators from "antifa." Steve Milloy, who joined the Heartland board of directors in 2020, and who tweets under the handle @JunkScience, suggested in a media interview that the riot was a set-up by the "deep state," and that the police and military "just let this happen so that they could set President Trump up for this impeachment."[27]

Heartland Institute PR flacks are paid for their big mouths. So, they're not afraid to mix it up online as attack dogs for fossil fuels and, apparently, also for Trump. But corporate CEOs usually like to keep a lower profile when it comes to partisan politics and Big Oil CEOs were more circumspect about the Capitol insurrection. Some even denounced the violence. Yet nice words only mean something when accompanied by actions.

Big Oil has sent money to Heartland and other propaganda outlets for decades. Oil companies also supported science-denying politicians who allied themselves with Trump's effort to overturn Joe Biden's victory and then egged on the Capitol insurrectionists. Republican Senator Ted Cruz of Texas has dismissed climate science as a "religion," not a science. Senator Tommy Tuberville, Republican from Alabama, has said that greenhouse pollution from burning fossil fuels has any effect on global temperature rise because God controls the climate. And Kansas Republican Senator Roger Marshall says he's "not sure that there is even climate change." All three are big beneficiaries of donations from fossil fuel companies. And the list goes on.

The man who provided so much funding to these senators and other Republican legislators who denied climate science and supported Trump, oilman and top conservative political donor Charles Koch, seems to be hoping that the public will forgive, if not quite forget.

Just after Trump was defeated by Biden in November 2020, Koch went on a tour to promote his new book on small government and apologized for his part in fomenting partisanship.

"Boy, did we screw up!" Koch writes in his *Believe in People: Bottom-Up Solutions for a Top-Down World*. "What a mess!" Koch made this apology after decades of paying front groups to oppose solutions not only to the climate crisis but to many other problems faced by ordinary Americans. As one journalist suggested, Koch deserved "the award for most destructive influence on modern American political life…He and his late brother, David, are the hard-right, libertarian gazillionaires behind almost every cruel, malign, extreme, divisive campaign that the GOP has run on in recent decades. Anti-regulation. Check. Anti-union. Check. Anti-healthcare. Check. Anti-environment. Check."[28]

Marcella Mulholland of Data for Progress suggests that Koch donations to Republicans who wound up challenging the electoral college certification of Joe Biden's election helped fuel the Capitol insurrection:

Koch Industries contributed $708,500 to Electoral College objectors through its PAC during the 2020 campaign cycle. Kansas' three Republican representatives all voted to overturn the will of the American people by voting against certifying the

2020 presidential results, as did Sen. Roger Marshall. The three objectors in the House all received $10,000 donations from Koch's PAC.

By putting campaign cash into the coffers of the Sedition Caucus, Koch Industries has rewarded and fueled their attacks on Democracy. What happened on Jan. 6 was the violent culmination of years of Republicans' undermining elections and spurring on dangerous, bigoted and violent rhetoric.[29]

After the attack on the Capitol, the Koch network promised to reevaluate its political donations. "Lawmakers' actions leading up to and during last week's insurrection will weigh heavy in our evaluation of future support. And we will continue to look for ways to support those policymakers who reject the politics of division and work together to move our country forward," said Emily Seidel, CEO of Americans for Prosperity and senior adviser to AFP Action, the group's super PAC, in a statement to POLITICO.

Only the future will tell if Koch donations stop flowing to supporters of Trump's attempted self-coup.

Meanwhile, other oil powers tried to distance themselves from Trump once he had become politically toxic after the January 6 insurrection. But their nice words in the present were inadequate to compensate for their dark deeds in the past. After law enforcement and National Guard troops had cleared the pro-Trump mob from the U.S. Capitol building, Chevron tweeted a message calling for a peaceful transition to the new Biden administration: "The violence in Washington, D.C. tarnishes a two-century tradition of respect for the rule of law. We look forward to engaging with President-Elect Biden and his administration to move the nation forward." Activists quickly pointed out that this sentiment rang offensively hollow.

Never mind that during his campaign, Biden promised that if he was elected, he would support climate-related litigation against Chevron and other oil companies filed by 20 U.S. states and localities.[30] Just as Lincoln was not the president favored by the Slave Power in 1860, so Biden was not the president that Big Oil wanted in 2020. Chevron and the rest of Big Oil gave as much support to Trump and asked for as many handouts from Trump's administration as the Koch network did. In 2019-2020 alone, Chevron made nearly a million dollars in donations to political action committees of federal candidates, with 85 percent going to Republicans including Trump along with

Senators Ted Cruz and Josh Hawley who led the effort in Congress to overturn Biden's election. Over the four years of Trump's administration, Chevron made more than $1.8 million in donations to federal candidate PACs, overwhelmingly to Republicans.[31]

"It's absurd for Chevron to pretend that it's not complicit in upending the peaceful transfer of power. Fossil fuel money has poisoned democracy by stalling climate legislation and making climate denial the official position of the Republican party. But it has also entrenched demagogues like Senators Ted Cruz and Josh Hawley, and the corrosive effect of Chevron's investment in them was on full display [on January 6]," said Brian Kahn of Earther.[32]

In the end, Big Oil got a big payback in exchange for supporting Donald Trump. "The Trump administration spent its lifecycle swindling the public for Big Oil: defending the industry in climate lawsuits, passing out drilling permits like candy, knocking down health and safety regulations during a pandemic, and more," according to ExxonKnews. "It's poetic that the closing act of the Trump Department of Justice, on his last day in office, would be to fight for polluters over people in the nation's highest court," arguing a case before the Supreme Court against a local government trying to hold Shell and other oil companies responsible for pollution.[33]

Big Oil's Promises to Help Contradicted by their Plans to Hurt

You might think that since climate change has gotten so scary on the one hand and that renewables have gotten so cheap on the other, the oil industry would recognize that the market would reject its dangerous product in favor of a safe alternative. It would make sense that fossil fuel companies would start planning to wind down their operations. But if you thought that, you'd be wrong.

Big Oil knows that their reign of error will end someday. But they don't care that for the climate, the end of oil needs to come sooner rather than later. If society allows them to keep operating, dirty energy companies won't shut down on their own. They'll fight to stay open as long as they can. That's why fossil fuel companies will have to be forced to shut down. "The era of global oil demand growth will come to an end in the next decade," said Fatih Birol, executive director of the

International Energy Agency. "But without a large shift in government policies, there is no sign of a rapid decline."[34]

Absent government action, fossil fuel CEOs have no intention of liquidating their own companies or even phasing out their production of dirty fuels and transitioning into clean energy, no matter what reassuring things they might say in their climate plans. The truth is the opposite. Major investor-owned petroleum companies are planning to keep oil and gas production steady, or even find ways in which they can continue to expand.

Big Oil clearly agrees with the slaveholders of the Old South: If you're not growing, then you're dying.

Later in this chapter we'll see how, even as they have stopped denying climate science and have even started to promise to cut their greenhouse pollution, fossil fuel companies are still developing programs to expand production of oil, gas, and coal.

Meanwhile, it will put the cruel expansion plans of a heartless industry into context if we look back into history at cotton planters' plans to expand even in the face of overwhelming consensus across the country and in the Atlantic world that enslaved labor had no place in the "modern" world of the mid-nineteenth century.

From our viewpoint in the twenty-first century, chattel slavery seems to be both so morally bad and so economically uncompetitive compared to free labor running the machines of the Industrial Revolution. But as we've seen, big slaveholders didn't agree. They claimed the opposite—that slavery was an institution so benign morally and with such a bright future economically that slave-powered agriculture deserved to expand to new areas. Right up until the eve of the Civil War, slaveholders thought the U.S. government would continue to help slavery grow, as it had done for decades.

The end of that deal with the election of Lincoln was exactly why the Slave Power insisted on leaving the United States and trying to start a new country that wouldn't get in the way of slavery's growth and instead would actively help slavery to expand.

For years, oil companies also insisted on their right to keep growing as long as there was a market for oil and gas. But as we've seen, now they've started to claim that they're done with unrestrained growth. They've even announced plans to reduce their greenhouse pollution. A few companies have promised to go further. In August 2020 BP became the first major oil company with plans to sell fewer fossil fuels in the short term, pledging to cut oil and gas production by 40 percent by

2030. That sounds like real progress, because such reductions will be necessary to reduce global greenhouse gas emissions. The company also announced plans to boost its investments in "low carbon" energy technologies including bioenergy, hydrogen, and carbon capture and storage, but admits that the bulk of its business will still be in oil and gas.[35]

BP has promised more than most oil companies, and Greenpeace says that it's a "necessary and encouraging start." Yet, there's reason for skepticism. First, there are obvious problems with the "low carbon" technologies in which BP plans to invest. None of them are as clean or as proven as solar and wind power.

Then, there's the company's track record. Just a few years earlier, BP also announced that it was going into clean energy, but without success. After rebranding themselves as a supposedly clean energy company under the name "Beyond Petroleum," BP quietly divested itself of its wind power division in April 2013. This followed the company's exit two years earlier from making solar panels after 40 years in that business. After its failed attempt to transition from fossil fuels to renewables back in 2013, BP told investors that it was just going to go back to fossil fuels and planned to become a "more focused oil and gas company and re-position the company for sustainable growth into the future."[36]

Clearly, all energy isn't created equal both from a pollution standpoint and from a business standpoint. BP's experience poking holes in the ground for oil and gas to squirt out didn't translate well into making photovoltaic panels and parts for wind turbines. Who's to say that the company will have any more success on "low carbon" energy this time than it did before? It's also unclear how serious the company is about producing less oil and gas in the future. But at least BP has promised to try. Shell and Total both also pledged in 2020 to bring their net emissions to zero by 2050.

Exxon, by contrast, hasn't promised anything close. Their CEO Darren Woods dismissed other oil companies' release of carbon-reduction plans as a "beauty competition" that was all appearance without substance. He's right that other oil companies have promised very little and are not likely to reach even the low bar they've set for themselves. But that's a poor excuse for Exxon to promise even less.

When you look at its climate pledge, it turns out that Exxon hasn't promised to reduce its overall climate pollution at all. It's just said it will try to be more efficient about polluting.

First, Exxon has explicitly said that it's not responsible for the climate pollution produced when customers use the company's product by burning oil and gas. That's the responsibility of end users, like car drivers and homeowners. Well, that's like a drug dealer saying that he's not responsible for the suffering of addicts. They make and sell a dangerous product, but they're offloading the responsibility to stop the harm onto the victims of their dangerous business.

Second, Exxon's own emission reduction plan put out in 2020 promises to cut the greenhouse gas "intensity" of its upstream operations—the part of its business that involves finding and extracting oil and gas—15 to 20 percent by 2025, compared with 2016 levels. This doesn't mean that the company has promised to cut its overall emissions from 15 to 20 percent. The company has only said that it will try to produce each barrel of oil or unit of methane gas with fewer emissions. But if it produces more oil and gas overall, then its emissions will rise.

"Exxon plans to up its production by 1 million barrels per day over the next 5 years," said Andrew Grant, head of oil, gas, and mining research at the financial think tank Carbon Tracker. "Reducing a minority of its lifecycle emissions by a small sliver is the thinnest of fig leaves for a big increase in overall emissions and a bet on continued business as usual."[37]

We should distrust any promises, even weak ones, made by corporations that have spread lies about climate science for 30 years or more, along with lies about their pollution for more than a century. When you look at their record, the evidence is that oil companies won't really try very hard to cut their emissions. And they certainly aren't likely to show much enthusiasm for selling less oil and gas.

Instead, while claiming that they're part of the climate solution, oil companies may speed up their expansion of fossil fuels. As Oxford University's Dieter Helm has warned, since public opinion seems to be finally forcing governments to take serious action on climate, oil giants are planning a final fossil fuel "harvest":

If we were serious about addressing climate change, we would leave some oil in the ground, so there is a scramble among big oil companies to make sure their assets are not the ones left stranded. Their answer is to pump as much as they can, while they still can, to keep delivering shareholder dividends. But the problem for the rest of us is that they are going to produce far more oil and gas than is consistent with the Paris agreement.[38]

Exxon's plans to squeeze more oil and gas out of the Permian Basin in Texas call for the company to produce more oil from a single U.S. state than the Kingdom of Saudi Arabia by 2030. At the same time, Exxon hopes to keep production high in oilfields across the globe. Meanwhile, other companies have already started or are planning new projects in north-west Argentina, in Kazakhstan's Kashagan oilfield, in the Yamal peninsula in Siberia and in the Barents Sea, as well as in Canada. In other words, their plans are to grow or die. Of course, while they grow, we die.[39]

In the end, there may be little difference between oil companies that have pledged "net-zero" carbon emissions and those that haven't. Net-zero may be just greenwashing whenever it comes from an oil company. Aside from a century-long record of big lies about the impact of their business on people and the environment, a good reason not to trust oil majors to lead the way on climate is that their CEOs and top executives are personally invested in the status quo, according to authors of a report on oil company net-zero plans published in *Science*.

"These companies cannot be relied upon to decarbonize at the speed and scale needed to align their emissions with a 1.5°C pathway because the senior executives and directors have annual compensation packages worth millions of dollars," according to authors Richard Heede of Climate Accountability Institute and Dario Kenner of the University of Sussex. "They are unlikely to do this because it could put the company out of business and jeopardize their personal wealth."[40]

The big problem is that fossil fuel companies still incentivize executives to produce more, not less, dirty fuel. Oil majors like Exxon and Shell tie executive compensation to metrics such as oil and gas production, the value of reserves on the books, and discovering and replacing oil reserves. In addition, across corporate America, the trend in recent years has been to pay executives less in direct salaries and more in stocks and other market performance measurements. As a form of deferred compensation, the value of those may not be available to CEOs and top executives until some point in the future. If the price of their company's stock falls, their future compensation could be worth much less.

"Slowing down the low-carbon transition is profitable for the boards of these Carbon Majors," the report authors wrote. "The boards are not prepared to take actions that would support the global shift from fossil fuels because...their personal wealth tied up in their company's fossil fuel production would be negatively impacted."

This explains how oil companies can announce plans to achieve net-zero emissions that seem to contradict those same companies' plans to drill more oil and more places: the net-zero plans are full of lies. The truth is that they plan to keep producing fossil fuels as fast as they can for as long as society will let them get away with it.

Oil's Foreign Policy

Foreign policy, both official and unofficial, also helps major oil companies do business around the world. In the antebellum years, when slaveowners controlled the U.S. State Department, it was official foreign policy to help protect slavery abroad as just another way to protect slavery in the American South. One reason was that slavery did offer real financial benefits to the economy, though at a high price in terms of cruelty.

Fossil fuels too, though dangerous to local communities located near oilfields and refineries when produced and deadly to the climate when burned, also helped build the economy and make the United States the world's sole superpower.

Just as slavery helped make America rich and powerful before the Civil War, so oil helped make us wealthy and strong afterwards. Fossil fuels even helped defeat the Confederacy: Coal ran the industry that helped the North outproduce the South in rifles and bullets, cannons and shot, locomotives and warships. Even the brand-new oil industry alone generated $8 million in tax revenue for the U.S. war effort.

Three decades later, coal again powered the fleet that helped capture Cuba and the Philippines in the Spanish-American War. After Teddy Roosevelt switched that same fleet over to oil, petroleum helped America prevail in both World Wars. An American embargo on oil exports to Japan was one of the reasons that leaders in Tokyo decided to attack the naval base at Pearl Harbor on December 7, 1941. With the U.S. Pacific fleet out of the way, the Japanese could seek for alternate sources of oil by conquering Indonesia from its colonial master at the time, the Dutch, without any worry of American interference.

Ever since World War II, oil companies have exerted an oversized influence over U.S. foreign policy, making the State Department into a tool of oil diplomacy and the military its enforcer. As armed conflict with the Germans and Japanese was ending, Franklin Roosevelt made a deal with King Saud that in exchange for supplying the United States with oil, our military would provide protection to Saudi Arabia. President Jimmy Carter pronounced his own "Carter Doctrine" to inform the rest of the world that any threat to America's supplies of oil abroad would be taken as an act of war.

But the most famous instances of armed conflicts to protect foreign oil supplies were the two Iraq Wars, together, among the longest running conflicts in American history. In addition, since the Cold War, the Navy has patrolled the Strait of Hormuz at the entrance to the Persian Gulf to ensure a steady flow of tankers with oil from Saudi Arabia, Iraq, and other Gulf states to world markets.

Though all administrations since Lincoln's have been friendly to the oil industry, the two most recent Republican administrations were the most oil-friendly presidencies in U.S. history. George W. Bush came from an oil clan in Texas and started the second Iraq War as a follow up to the first one that his father had fought when the senior President Bush occupied the White House. The next Republican president was Donald Trump.

And though Trump had little connection to oil before he was elected, appointing Exxon CEO Rex Tillerson his secretary of state helped to elevate Big Oil's global interests into official U.S. foreign policy as never before. As CEO of Exxon, Tillerson became famous for making deals with Russia, and Vladimir Putin liked Tillerson so much that he awarded the Texas oilman the Russian Order of Friendship in 2013. It turns out that some of Tillerson's Russian deals were illegal, violating U.S. sanctions put in place to punish Putin's invasion of Ukraine in 2014 and earning Exxon a $2 million fine from the Treasury Department. And once confirmed, Tillerson served only 15 months as secretary of state before Trump sent him packing.[41]

So ended Big Oil's most official chance to run American foreign policy from the inside. But the industry still had plenty of access to decision-makers in the Trump White House, and they continued to get plenty of gifts from the administration that they would not have gotten under Hillary Clinton and began to lose under Biden.

Meanwhile, major oil companies continued to make plans invest in new drilling. As of 2020, the world's 50 largest oil companies were planning to flood world markets with an additional 7 million barrels per day over the coming decade, despite scientists saying that burning all this oil will generate enough climate emissions to push atmospheric heating beyond safe levels. Shell and Exxon are among the most aggressive companies, planning to produce enough oil and gas to increase climate pollution by 35 percent between 2018 and 2030, a sharper rise than over the previous 12 years. The planned jump in production is almost the opposite of the 45 percent reduction in carbon emissions by 2030 that scientists say is necessary to have any chance of holding global heating at a relatively safe level of 1.5 degrees Celsius.

At least 14 of the 20 biggest historical carbon producers aim to pump out more hydrocarbons in 2030 than in they did in 2018. "Rather than planning an orderly decline in production, they are doubling down and acting like there is no climate crisis. This presents us with a simple choice: shut them down or face extreme climate disruption," explains Lorne Stockman of Oil Change International, which monitors oil companies.[42]

Exxon even has its own foreign policy, much as slaveholders did in the nineteenth century. For example, in late 2011, the company was powerful enough to bully the government of Iraq to let them drill in Kurdish territory, despite Bagdad's objection that this could help lead to the breakup of its fragile nation. But with Iraq's revenue from oil sales that year at $80 billion and Exxon's worldwide revenue at $430 billion, Exxon had enough clout to push around Iraq's government.

Since then, Exxon's fortunes have greatly declined, as we've seen. Yet, even after so much pummeling on the international crude market and on the stock market alike, Exxon is still committed to producing more oil and gas, not less.

"Some believe the dramatic drop in demand resulting from coronavirus reflects an accelerating response to the risk of climate change, and suggest that our industry won't recover," CEO Darren Woods told the company's 75,000 staff in late 2020. "But as we look closely at the facts and the various expert assessments, we conclude that the needs of society will drive more energy use in the years ahead—and an ongoing need for the products we produce."[43]

What will they do with all this oil? Demand for liquid fuels may have peaked in 2020 worldwide. Demand may be on an irreversible decline in North America and Europe, but until they get enough renewable energy, developing countries will still want more oil. And Exxon and its peers hope to still sell lots of natural gas, falsely pushing it as a "bridge fuel," to help developing economies transition from fossil fuels to clean energy.

Dirty energy companies especially want to grow abroad in places like sub-Saharan Africa, where 580 million people lacked electricity as of 2020. That could get worse in the future as poverty levels climb because of Covid and those who had electric service lose it, reversing years of progress (thus, the silly claim by the industry that there's a "moral case for fossil fuels").[44]

Back home, oil producers are placing their bets on plastic. Though accounting for only a small part of petroleum use now, the industry predicts that plastics will be the biggest (and perhaps only) source of new demand for oil in the coming decades. That's a bad bet for four big reasons, according to a 2020 report from Carbon Tracker: High carbon emissions, high externalized costs in terms of other pollution, exceptional wastefulness where 36 percent of plastic is single use, and increasing opposition from the public. Global plastic demand is projected to peak in 2030.[45]

The oil industry is also promoting hydrogen as an alternative to liquid fuels to power transportation, as a feedstock for chemicals, and to store energy from intermittent renewables such as solar and wind power. But hydrogen may just be a sneaky way to keep gas pipelines in use—which turns out to be unsafe because hydrogen will damage pipeline walls and lead to leaks that can cause explosions—while creating demand for more methane to make all that hydrogen.

As with so many allegedly clean energy technologies pushed by oil companies, when it comes to hydrogen, it turns out that the devil is in the details.

Hydrogen is not an energy source, but merely an energy carrier. So, producing hydrogen from water or from other materials doesn't generate new energy but merely creates an alternative to liquid fuels like gasoline and diesel fuel. It takes more energy to make hydrogen than the hydrogen will provide when it's burned. Nonetheless, the European Union's plan to get to net zero climate pollution by 2050 relies heavily on hydrogen to reach its goals.

Batteries can store energy better than hydrogen, and batteries are getting cheaper all the time. The cost of batteries for use on the electrical grid dropped 70 percent over just a three-year period from 2017 to 2020.[46] Better batteries have helped vehicles from cars and trucks to trains start to transition to electricity at a fast clip as well. Whether large ships and airplanes can make the switch to electricity remains an open question, so there could be some limited use for hydrogen in transportation. The big problem is that most ways to make hydrogen are neither clean nor renewable. Zero-carbon or "green" hydrogen, made from the electrolysis of water powered by solar or wind energy, is the only clean kind. But no surprise, it's the most difficult and costly kind of hydrogen to produce.

In fact, green hydrogen is so expensive as to make it impractical in the foreseeable future. But that's OK for the industry since they don't plan to sell green hydrogen anytime soon anyway. The promise of 100 percent renewable hydrogen is just a way to fool the public and provide advertising cover to promote hydrogen in general, which will certainly be dirtier and will likely be either of the "grey" variety made from methane gas or the "blue" variety, made the same way, but with the carbon emissions captured and stored, adding significant additional expense.

Either way, "blue" or "green" hydrogen together made up less than one percent of the hydrogen market in Europe in 2020, while dirty "grey" hydrogen made from methane with no emissions captured provided more than 90 percent of Europe's hydrogen.

"Ads from big oil companies promoting green hydrogen are misleading the public about the industry plans for fossil fuel based hydrogen because green hydrogen is currently prohibitively expensive and nowhere near being able to scale up to be a major contributor to the near-term energy markets," writes climate journalist Justin Mikulka. He points out that if hydrogen were really a viable alternative to oil and methane gas, then oil and gas producers would not be touting hydrogen, as it would destroy demand for their core fossil fuel products.[47]

Oil and gas producers want to keep pumping out oil and gas. But if we let them have their way, the climate will be cooked.

So far, oil and gas companies are still winning. While they've stopped saying defiant things and started to claim that they're ready to cooperate on cutting climate pollution, their actions and their business plans really say the opposite. Despite calming words, it's clear that fossil fuel companies are not serious about climate but remain committed to producing and selling dirty energy if they can get away with it. Given the historical momentum and the massive political power of Big Oil, it won't be easy to abolish fossil fuels, even gradually. Either a bigger stick or a bigger carrot will be needed to make any progress towards shutting down fossil fuels.

Whatever reassuring things oil companies may say about being part of the solution to bring climate pollution down to safe levels in the next ten years or even by 2050, history shows that the Oil Power is unlikely to give up trillions of dollars of wealth without a fight.

That's just how it was for slaveowners in the nineteenth century. Asking them nicely wasn't enough to get big planters to give up their slaves. It took a bloody civil war to bring abolition to the United States.

But another major slaveholding nation, Great Britain, managed to emancipate its enslaved people largely peacefully. The way they did it could be a model to abolish oil in time to make a difference to save the climate. Or it could suggest the dangers of compromises with the Oil Power like phony net-zero emissions pledges that don't cut climate pollution but instead just give oil companies public relations cover to keep pumping and polluting for years to come.

6

Abolition without War

The cost of damage to property and lost productivity in the Civil War
turns out to be between 12 and 18 times more expensive than the
compensation that was agreed in Britain.

—Richard Barker, investment analyst, Iona Capital, London

If compensation be demanded as an act of justice to the slave holder,
in the event of the liberation of his slaves; let justice take her free,
impartial course; let compensation be made in the first place where it
is most due, let compensation be first made to the *slave*, for his
uncompensated years of labor, degradation and suffering.

—Elizabeth Heyrick, British abolitionist, 1824[1]

Three decades before the United States abolished slavery after
the bloodiest war in its history, the British Empire abolished
slavery peacefully by paying slaveowners compensation for the
"property" value of the people they held in bondage. As morally
repugnant as it was in the 1830s to reward the very people whom
abolitionists of the day referred to as "man stealers," paying enslavers
and their investors to release their property rights in humans may have
helped lead to a largely peaceful and orderly emancipation of enslaved
people in the 1830s in the British Empire.

Today, various market-based climate solutions purport to allow
fossil fuel companies to continue to do business in some form while
society reduces its overall pollution from fossil fuels. From a carbon tax

that would make fossil fuels more expensive and thus reduce consumer demand for them; to capturing carbon from smokestacks and storing it underground; to offsetting pollution by putting in trees or putting up solar panels; to "cap-and-trade" programs that offer tradeable emissions credits that incentivize companies to help each other to cut pollution, ideas abound that would let oil, gas, and coal companies keep production going while at the same time claiming to reach "net zero" climate emissions by 2050 or some other date in the future.

Some of these ideas, like a well-designed carbon tax, might help cut pollution. But other schemes supported by the oil industry offer doubtful benefit to the climate. "It's naive of us to think that all of a sudden the oil and gas industry is going to put forward policies that are going to keep fossil fuels in the ground," according to Jim Walsh, senior energy policy analyst for environmental NGO Food and Water Watch.[2]

All these ideas represent a form of compensation to the industry that is supposed to help cut climate emissions in the short term and presumably, at some point, lead to gradual abolition of fossil fuels.

The proposal to compensate slaveowners in Britain's island colonies in the Caribbean to gradually emancipate their slaves was opposed by abolitionists in the 1820s and 1830s. Likewise, today many environmentalists criticize programs and ideas to encourage the Oil Power to gradually transition out of fossil fuels while still getting some value from its sunk investments in drilling rigs, refineries, pipelines, and massive earthmovers used at sites of mountaintop removal coal mining.

Except to keep workers employed, programs to let fossil fuel companies continue to operate seem unfair and unjust. Climate justice calls for helping the people who've suffered most from fossil fuel pollution, low-income frontline communities and especially people of color, not for ways to help those rich and heartless companies who made the mess in the first place.

And do market-based programs even work to cut climate emissions? Or are they just deceptive schemes based on clever accounting tricks to get society off the industry's back for a few more decades?

Since oil, gas, and coal producers are some of the biggest liars in the history of business, skepticism should be the first reaction to any idea to allegedly cut climate pollution that fossil fuel companies support. But paying an oppressive industry to walk away from its oppressive practices may have helped end slavery in the nineteenth century.

Could paying fossil fuel companies to ramp down and eventually walk away from their dirty industry help to save the climate today? Given the relatively smooth way that Britain abolished slavery versus the bloodbath that it took to end slavery in the United States, it's worth having a look at Britain's model for gradual and compensated emancipation. Does it offer any lessons for abolishing oil more quickly?

To start with, it wasn't perfect. The problems with paying British slaveowners to walk away from the fight over abolition were as obvious at the time as they are today.

No compensation, reparations, or other financial help was given to the newly freed people, either to recognize lives of unpaid work or to ease their transition into the economy. And while payments to freed people were nonexistent, the payout to slave masters was massive, so high that British taxpayers couldn't retire the debt until 2015. When the British Treasury tweeted about the payoff in 2018, social and racial justice advocates, the modern-day political descendants of abolitionists, strongly objected.

"Generations of Britons have been implicated in a legacy of financial support for one of the world's most egregious crimes against humanity," wrote historian and activist Kris Manjapra. Meanwhile, leaders of Caribbean nations, former colonies of Britain, had already renewed calls for slavery reparations.[3]

Also, British abolition was not entirely peaceful. Emancipating slaves without a civil war among white Britons came only at the end of two centuries of violence by slave traders and slaveowners against captured Africans and their descendants. White people back in the British Isles weren't affected by the violence of slavery. But in the West Indies it was everyday business for planters and overseers to force enslaved people to plant, harvest, and process sugarcane in the blazing sun from dawn till dusk without pay and without control over their own bodies. Slavery was nothing without violence. The hoes, machetes, and boiling pots of sugar production were less important tools than the whips, chains, and shackles that compelled people to work against their will.

Oppressed people in Britain's Caribbean colonies found different ways to resist slavery, including launching more than a dozen major slave rebellions from the seventeenth through the nineteenth centuries. Deaths in slave rebellions over three centuries ran into the hundreds or thousands.

Ultimately, it is impossible to quantify suffering and death from centuries of slavery. But in terms of open warfare, in a four-year period the United States had to suffer more than 620,000 soldiers killed in the Civil War before it was able to end slavery with the Thirteenth Amendment in 1865, a cost in lives which Britain managed to avoid. Financially, it was nearly twenty times more expensive for America to abolish slavery by war than for Britain to free slaves by paying compensation to slaveowners.

The Politics of Slavery in Britain vs. the United States

Slavery in the British Empire was easier to unravel than in the American South for one primary reason: Slavery wasn't as important to the British as it was to the Americans. In the United States, slavery made the South into the richest region of the country. In 1860, at its peak, slave-grown cotton was America's leading export product. The slave property of the South was worth more than all the railroads and factories of the North, combined. Americans enslaved five times more people than Britons held captive on colonial plantations.

Another thing that made slavery in America so hard to end was that it was found right at home in fields, houses, and workshops across fifteen Southern and border slave states. The nearness of slavery required the United States to develop a rigid system of racial hierarchy to separate the free from the enslaved, a caste system that became deeply and tragically embedded in all aspects of American culture. More than a century and a half after the end of slavery in the U.S., slavery's cultural legacies of white supremacy and racism remain difficult to dislodge today.

By contrast, British slaves labored not at home but far away in island colonies in the Caribbean, meaning that the British sense of national identity never came to rely on race the way it did in the American South and even across the North. Through most of the period of West Indian slavery, slaves at home in Britain were a rarity. If brought to Britain, their legal status as property was uncertain, and most British people in the eighteenth and early nineteenth centuries never set eyes on a single enslaved person or even a free Black person in their whole lives.

Related to its remote location were other factors that made slavery in the British Empire less difficult to end than in the United States. In slave states, governors and members of state legislatures were slaveholders. But the control of the Slave Power went beyond the states where slavery was legal through outsized representation of slaveowners in the federal government.

We've already seen that two-thirds of the presidents before Lincoln owned slaves and so did many other holders of high federal offices, from cabinet secretaries to members of Congress to Supreme Court justices. Also, even presidents and politicians from free states were often sympathetic to slavery for economic reasons. The Slave Power didn't just lobby the government in Washington. Most of the time before Lincoln took office in 1861, the Slave Power ran the government in Washington. By contrast, in Britain, the power of slaveholders over the national government was more limited.

Slaveholders certainly had influence over government in the United Kingdom, of course. The whole industry of Caribbean sugar planters, shipping interests, and textile producers cultivated powerful friends at court, where the royal family generally was hostile to preachy abolitionists who pressured the world's largest economy to put morals ahead of money.

West Indian planters, many of them absentee landlords who lived in London, also had many friends in Parliament, including members of that legislature who were slaveowners themselves. And for most of the period before abolition, property requirements denied working-class Britons the vote. So, the electorate was limited largely to wealthier people, those most likely to have investments in the West Indies.

Added up, slaveowners certainly had big influence over the British government. Still, that influence did not rise to the level of control exercised by the slave-owning class in the U.S. That's why in Britain it was easier than in the U.S. for top government officials and business leaders to ally with abolitionists and work to pass legislation to restrict and ultimately end slavery.

Then there was the money. Slavery in the American South before the Civil War was highly profitable right up to the end because it dominated the market for the world's most widely traded commodity, cotton. It was different in the British Empire whose most lucrative colonies were found in the Caribbean and produced a different commodity crop, sugar. By the late eighteenth century, plantation agriculture in the British West Indies was in economic decline as competing sources for sugar came onto the market from Spanish colonies and Brazil. When British abolitionists recruited 400,000 tea drinkers to boycott West Indian sugar in the early 1790s, retailers soon responded by sourcing "slave-free" sugar imported from British colonies in India that consumers believed was produced in less harsh conditions.

Finally, because American slavery was based at home in slave states, the main issue that slaveowners used to oppose demands by abolitionists that the federal government abolish slavery was states' rights, an American constitutional principle which has remained to the present day a potent argument to thwart national progress on a range of problems from civil rights to abortion to healthcare.

Since British slavery happened overseas, it was not associated closely with any region of the country back home. And since the places where British slavery took place were all colonies answerable to the Crown and Parliament where local governors and colonial legislatures enjoyed limited autonomy, local rights were not much of an argument to push back against interference in local slavery by the national government.

Instead, the main political issue in British abolition was not states' rights but property rights. That's why providing financial compensation to slaveowners ultimately proved to be the magic formula to untie the Gordian knot of peacefully ending slavery. By contrast, money wasn't enough to buy off American slaveowners, as we saw previously when, during the Civil War, even slave states like Delaware and Maryland that remined in the Union rejected Lincoln's offer of compensated emancipation.

Overall, Britain's overseas slaveowners had less power than America's domestic Slave Power and were thus easier to bring to the negotiating table when public support for abolition reached a high enough level. This may make the example of British abolition more relevant than the story of the American Civil War to abolishing oil today.

Britain did not require a civil war to end slavery in its overseas colonies but still, abolition didn't come either quickly or easily. It took 50 years of campaigning by British abolitionists, along with help from world events, to convince West Indian planters that accepting compensated emancipation was a better deal than they'd likely get from continuing to fight to try to keep their slaves. The threat of violence also played a role.

The business of buying and selling slaves and producing sugar with slave labor in the Caribbean was very lucrative for a very long time.

"Portugal and Britain were the two most 'successful' slave-trading countries accounting for about 70 percent of all Africans transported to the Americas. Britain was the most dominant between 1640 and 1807 when the British slave trade was abolished. It is estimated that Britain transported 3.1 million Africans (of whom 2.7 million arrived) to the British colonies in the Caribbean, North and South America and to other countries," according to the British National Archives.[4]

Forcing slaves on Caribbean islands to perform the grueling work of planting, tending, and harvesting sugar cane, and then crushing the stalks and boiling them down into dry sugar and molasses that could be shipped to North America or back to Britain, paid handsomely.

In 1773, just before the American Revolution broke out, the value of imports from the single island colony of Jamaica, Britain's leading sugar producer, was five times greater than the value of products from all 13 colonies from Massachusetts to Georgia. Britain would fight hard to keep those North American colonies from gaining their independence, so you can imagine how much the British valued Jamaica.[5]

The big money in the transatlantic slave trade and in West Indian sugar production helped produce many of the same outcomes as the big money in Southern slavery did in the United States. We already saw how Southern slaveowners helped run American foreign policy during the first half of the nineteenth century, demanding foreign invasions to seize territory in Central America or the Caribbean that Southern leaders hoped could be carved up into more slave states in the future.

A few decades earlier, British slaveowners had hijacked Britain's foreign policy too, though with less success. Their crowning achievement was a failed invasion of the French colony of Haiti in the late eighteenth century.

After the outbreak of the French Revolution, the teachings of *liberté, egalité, fraternité* reached the enslaved people of France's largest sugar colony in the Caribbean, San Domingue, later known as Haiti. Inspired by revolutionary ideology and taking the opportunity of the political turmoil in France, slaves launched a revolt in the island colony, starting with the "Night of Fire" on August 22, 1791, when they burned plantations and sugar-processing facilities. Under the generalship of Toussaint Louverture, a former slave, rebels routed French forces and gained control of much of the colony.

Other colonial powers including Britain feared that if the Haitian revolution were allowed to succeed, it would spark slave rebellions throughout the Caribbean.

To avert this threat, planters in Jamaica and other British Caribbean islands clamored for the British military to occupy Haiti and to return the self-emancipated slaves to bondage. So, when Britain went to war with Revolutionary France in 1793, at the urging of powerful West Indian planters, British troops invaded Haiti, capturing a large part of the insurgent colony, and occupying it for five years. But the Haitians fought the invaders as well as they had fought the French. In the face of fierce local resistance, British forces failed to pacify the insurgent colony.

Meanwhile, the revolutionary French tried to woo back the insurgent Haitians. France abolished slavery in all its colonies in 1794 (a decree that, sadly, would be rescinded by Napoleon a few years later). But this failed to rally Haitians in support of their former colonial masters in Paris even as they fought against the British invaders sent by London.

The Haitians fought bravely to keep their freedom and push the British back into the sea. The world was stunned to see that armies of self-liberated Black Haitians could successfully fight off trained and well supplied redcoats of King George III. After suffering tens of thousands of deaths from fighting and disease, British forces withdrew from Haiti in disgrace. The world's leading military power had failed in its ignoble mission to force a hard-fighting and freedom-loving people back into chains.

Meanwhile, events in the Caribbean energized the abolition movement back home in Britain. Abolition's champion in Parliament, William Wilberforce, submitted a bill in 1792 to end Britain's involvement in the transatlantic slave trade. Though it would not end slavery in Britain's empire, removing Britain from the cruel and dangerous business of buying captives in Africa and shipping them to buyers in the Americas would drive a big nail into the coffin of slavery.

Ending the trade in slaves was politically more practical at the time than ending slavery itself since slaveowners didn't fight against it as hard. Even if they couldn't add new enslaved people from Africa, West Indian planters could keep the slaves they already had.

Surprisingly, some slaveowners saw an upside: Ending the British slave trade would put pressure on other nations including Spain and Portugal to prohibit the slave trade as well. This in turn would restrict labor supplies to those nations' Caribbean colonies which competed in the sugar market with British planters. Starting from a larger enslaved population, British planters hoped to grow the size of their slave workforce through natural increase faster than the Spanish and Portuguese could in their sugar-growing colonies.

It would indeed prove easier to abolish the trade in slaves than to end slavery overall. But fighting the war against France put all abolition efforts on hold. It still took 15 years for Wilberforce's bill taking British ships out of the Atlantic slave trade to pass, which it finally did in 1807, going into effect on January 1, 1808. After stopping the capture and shipment of enslaved Africans, it would take another 14 years to pass legislation to end slavery itself in the British Empire.

The strategy of winning partial victories proved effective for British abolitionists in the early nineteenth century and offers a powerful lesson to the climate movement today. Measures like a carbon tax or bans on drilling on federal lands or requiring warning labels on gasoline pumps won't abolish fossil fuels right away. These restrictions or discouragements on fossil fuel use may not even be sufficient to reduce climate pollution as quickly as scientists recommend.

That's why some activists think partial victories are harmful, arguing that they give fossil fuel companies a license to keep polluting for years to come. All-or-nothing seems to be the approach of some people in the climate movement. Yet, if they don't distract from bigger gains, partial victories have great value politically in setting a precedent for bigger wins later on as well as getting the public used to thinking about an industry as something that really is starting to go away.

Of course, fake solutions like turning oil into more plastics or using natural gas to make hydrogen are a different story. Anything that creates new uses for dirty energy should be strongly resisted. But taxes, fees, and regulations that genuinely reduce the production and use of fossil fuels in the future are squarely in the tradition of successful slavery abolition in the past.

Ending the international trade in slaves did not end slavery in British colonies any more than it did in the American South when the U.S. ended its own international slave trade in the same year, 1808. But stopping the supply of new slaves from Africa was a step towards liberating the enslaved people who were already in Britain's Caribbean colonies. Ending the slave trade was a way of putting an expiration date on slavery as a viable institution.

When it comes to oil, this same approach was tried once in the past in the United States. In 1975, after OPEC embargoed exports to the U.S. and other nations supporting Israel during the 1973 Arab-Israeli War, Congress responded with legislation directing the president to stop American producers from exporting most types of unrefined petroleum to foreign buyers.

The ban lasted until 2015, when the American Petroleum Institute and their allies succeeded in lobbying the federal government to lift the ban. If the crude oil export ban had remained in place, it would have reduced the markets for American oil producers and perhaps kept prices down. That would have cut into oil companies' revenues and profits. Worse financial performance in turn would have weakened the companies in general and reduced their influence over politics and their ability to fight against climate solutions.[6]

Inventing the Modern Political Advocacy Campaign

Successfully ending the transatlantic slave trade and then working to abolish slavery altogether took hard and creative campaigning by abolitionists who capitalized on events as they arose. In the process, British abolitionists invented many of the elements of a modern campaign on a political issue. It's worth examining what abolitionists in Britain did in the eighteenth and nineteenth centuries in some detail to get ideas for the climate movement now.

"Never doubt that a small group of thoughtful committed individuals can change the world. In fact, it's the only thing that ever has," advised Margaret Mead, in one of the most famous quotes in political activism. The beginning of the abolition movement in Britain proves Mead's point. It all started in the 1760s with a few scattered individuals who later got together as a very small group that would grow into a powerful movement.

The British did not invent the transatlantic slave trade, but since at least the seventeenth century British ships had been carrying slaves to the New World. Enslaved people started resisting since the first African was brought in chains aboard a European ship. Slaves staged at least a dozen armed rebellions on British Caribbean islands, bringing violent insurrection every few decades. Yet, in a world where overseas wars between Britain and other colonial powers like France and Spain broke out every twenty years or so, violence in the Caribbean failed to attract much attention back home in Britain.

And by the mid-eighteenth century, the trade in West Indian sugar was big business that generated big money for a few top merchants and landowners and good money for many other Britons. Only a handful of people in Britain at the time opposed slavery. One of them was attorney Granville Sharp.

In legal cases in the 1760s and 1770s, Sharp successfully defended Black men who had been brought to Britain by their masters and had escaped from being re-enslaved, including James Somerset. Decided by Chief Justice Lord Mansfield, the famous *Somerset* case wound up establishing a legal precedent that prevented colonial slavery from taking root on British soil. Essentially declaring that once a slave set foot in Britain, he or she became instantly free, the decision provided a legal path for many slaves to seek their freedom.

Another famous case that Sharp argued in 1783 involved the captain of the slave ship *Zong* who threw 132 African captives overboard to collect insurance money. The ship owners' lawyer argued that the slaves were "property" and the court agreed, which prevented Sharp from bringing a case for murder but highlighted for abolitionists the inhumanity of the slave trade.

In the same way, using the courts to enforce environmental regulations has been a tactic of environmentalists in the United States since the passage of landmark clean air and water legislation in the 1970s. Non-profit legal groups such as the Environmental Defense Fund, EarthJustice (tagline: "Because the Earth Needs a Good Lawyer"), and the Southern Environmental Law Center have coordinated many legal challenges to hold polluters accountable. Their able legal work has won decisions forcing polluters to spend millions of dollars on clean-up efforts and civil and criminal fines.

Later, climate activists started filing legal cases seeking to hold fossil fuel companies responsible for the effects of climate pollution and demanding financial compensation. Recently, in the U.S., state and municipal governments have joined these suits, clearly scaring fossil fuel companies and their investors. For example, both New York State and New York City have sued oil companies for false advertising, claiming to be part of the solution on fighting climate heating when those companies were still doing business in a way to make pollution worse.[7]

Radical Activists Recruit Mainstream Allies

The British abolition movement expanded beyond the courtroom when Sharp attended a small meeting on May 22, 1787, held at a printing shop in London. Its agenda was to form a committee dedicated to ending the slave trade.

Though nine of the twelve participants were Quakers, the group elected Sharp and another Anglican, Thomas Clarkson, as president and vice president respectively of the new association. On both sides of the Atlantic, Quakers had taken the lead on agitating against slavery for years, as we saw in America in the story of the uncompromising and theatrical early abolitionist Benjamin Lay. But in Britain the impact of Quakers was limited because they were considered an outsider group in a nation whose religious life was dominated by the government-run Anglican Church.

Putting a couple of Anglicans at the head of their group gave abolitionists more credibility. It also provided access to government officials, primarily conservative Member of Parliament William Wilberforce, who Quakers could not reach alone. Once convinced that the slave trade was both morally wrong and bad for the future of the British economy, Wilberforce became abolition's main spokesman inside the British government.

Though their final goal was to end slavery altogether, the group chose strategically to fight first only on a preliminary goal, to end the transatlantic slave trade. The committee named itself accordingly: The Society for Effecting the Abolition of the Slave Trade. Aside from the zeal of its members, the new group's main asset was a printing press, a network to distribute pamphlets and books, and a growing number of influential allies recruited by Sharp and Clarkson.

There's a lesson in the alliance between Quakers and Anglicans for the modern climate movement about the need for more radical activists to form alliances with those in the center. British abolitionists also formed alliances across race. In Britain as in America, former slaves became leading abolitionists.

The way abolitionists highlighted the stories of enslaved people themselves foreshadowed today's alliance between mainstream environmental and climate groups, many of whom are white, with the movement for environmental and climate justice, focused on people of color.

Stories by people from "frontline" communities who have experienced the worst pollution make an important connection between the two kinds of pollution caused by Big Oil—the visible fouling of air, water, and land in local areas on the one hand and the invisible heating of the global climate on the other. In the same way, the British abolition movement shared the stories of Black people who had suffered in slavery and could describe the horrors of the slave trade first-hand. This made the evils of bondage experienced by slaves in the Caribbean more real to white people back home in Britain.

The most famous memoir by a formerly enslaved person, *The Interesting Narrative of the Life of Olaudah Equiano, or Gustavus Vassa the African,* quickly sold out its first edition of 700 copies. The book was reprinted in a dozen British editions during Equiano's lifetime and in the two decades after his death, and translated into German, Dutch, and Russian. Full of adventure and pathos, Equiano's autobiography gave the lie to claims by slaveowners that Africans were inferior to Europeans. That the author was an avowed Christian made the story relatable to his British readers in an age when the nation overwhelmingly shared the same Christian faith.

Equiano's story began when he was kidnapped from his home in the Niger Delta at age 11 and taken to the West Indies. Sold to an officer in the Royal Navy, Equiano served bravely on a warship in the Seven Years War. Then, after being sold again to a merchant, Equiano learned the business of trade.

While helping his master, Equiano bought and sold his own goods on the side. Eventually raising enough money to buy his freedom, he settled in London but continued to travel abroad, even joining an expedition to the Arctic in 1773. "It was difficult for those who read the book not to associate themselves with the African hero who was courageous, resourceful, literate, cultured, and Christian—all qualities that British people of that time admired and aspired to," explains Mike Kaye of Anti-Slavery International.[8]

Equiano and other freed slaves in Britain became leading abolitionists who toured the country giving speeches on the evils of the slave trade. Their testimony was especially useful to refute the public relations campaign of the slaveowners, coordinated under a lobbying group called the West India Committee, which might qualify as the world's first science-denying industry lobby. Levying a fee on planters for every barrel of sugar or rum or every bag of cotton imported, the proslavery business cartel amassed a war chest to pay for attorneys to oppose Granville Sharp in court, to put out publications extolling the benefits of the slave trade, and to lobby Parliament and the King to resist calls from abolitionists to restrict or regulate the slave trade.

"Just as the abolitionists were the prototype of modern citizen activism, so the West India lobby was the prototype of an industry under attack," writes historian Adam Hochschild. In much the same way, today the American Petroleum Institute lobbies for Big Oil and fights against government action to limit pollution from fossil fuels.

One of the West India Committee's first moves was to start a disinformation campaign to fool the public into thinking that slavery wasn't so bad. One proslavery writer, perhaps in jest, suggested change to wording, a kind of rebranding:

> The vulgar are influenced by names and titles. Instead of SLAVES, let the Negroes be called ASSISTANT-PLANTERS; and we shall not then hear such violent outcries against the slave-trace by pious divines, tender-hearted poetesses, and short-sighted politicians.[9]

Industry lobbies, whether then or now, always try to put the best face on how they make their money. Sometimes their PR spin is so rosy that it veers unintentionally into satire. Testimony given before the King's Privy Council by Robert Norris, a retired slave captain from Liverpool, was typical of the spin put out by the slave shippers' lobby. Norris claimed that captives held aboard slave ships sailing from Africa to the Americas "had sufficient room, sufficient air, and sufficient provisions. When upon deck, they make merry and amused themselves with dancing. As to mortality...it was trifling. In short, the voyage from Africa to the West Indies was one of the happiest periods of a Negro's life."[10]

In response to such claims, oft repeated in industry propaganda, Equiano narrated his own experience on board a slave ship. "The air soon became unfit for respiration, from a variety of loathsome smells, and brought on a sickness among the slaves, of which many died, thus falling victim to the improvident avarice, as I may call it, of their purchasers." Looking around the ship, Equiano saw the very opposite of the joy described by Captain Norris on the faces of his fellow Africans, "a multitude of Black people of every description chained together, every one of their countenances expressing dejection and sorrow."[11]

British abolitionists used other tactics that later became important to advocacy campaigns and can inspire effective communications about the climate threat and the value of climate solutions.

Thomas Clarkson rode 35,000 miles around Britain and went on board hundreds of slave ships docked at home ports between voyages to carry out interviews with seamen, merchants, and ships' surgeons to collect evidence on conditions in the Atlantic slave trade.

During his visits to slave ports in London, Bristol, Liverpool, and other shipping centers, Clarkson also amassed a collection of props to make that evidence tangible. Objects included iron shackles, thumbscrews, and a tool for force-feeding slaves on hunger strike. Clarkson brought these gruesome artifacts along to meetings with government officials to help them understand the barbarity of the slave trade.

Clarkson supplemented this grim collection with another more hopeful one. One of the first ships Clarkson boarded was not a slaver but a trader called the *Lively* whose hold was full of trade goods from Africa: carved ivory, gold jewelry, and woven cloth along with produce such as beeswax, palm oil, cotton, indigo, and peppers. Clarkson bought samples from this cargo, which formed the start of a collection that he kept in a large wooden box.

At abolition meetings and public lectures, Clarkson pulled out his box, and used the contents to demonstrate the skill of African artisans and the bounty of African growers. Then he extolled the lucrative possibilities for a more humane alternative to the slave trade that could benefit British shippers and merchants alike.

The objects in Clarkson's box were precursors to the portable solar panels or sleek electric cars that climate activists would later bring to marches and eco-fairs alike to show that the economy didn't have to be dirty to prosper. You could protect the climate and still make money.

The movement for clean energy has also learned to extend its sympathies to coal miners, oil rig employees, and other ordinary workers in fossil fuels. With health conditions such as black lung disease, coal miners are the most prominent example of workers who suffered while making a living in fossil fuel production. But workers on offshore oil platforms, in refineries, and in fracking operations also died in explosions or suffered ongoing health impacts. This has helped refute industry claims that fossil fuels offer good employment.

British abolitionists extended their sympathies beyond captive Africans to white sailors on slave ships. Through his interviews with sailors and by poring through official records, Clarkson also found that suffering was not limited to slaves kept chained in a ship's hold but also affected sailors on deck too. On average, 20 percent of each ship's crew died from disease or ill-treatment by officers during each voyage of a slave ship.

Once slaves were loaded in West Africa, a captain had an incentive to treat his crewmen badly, since contracts typically stated that sailors would only be paid if they made it back to port in Britain. When sailors deserted their ships or when captains discharged crew members in West Indian ports, it helped cut costs and increase the owners' profits. Clarkson's evidence disproved the popular belief that slave ships provided good jobs or good training for inexperienced seamen and showed how the slave trade hurt white people as well as Black ones.

Pictures Louder than Words

In 1788, Clarkson set out on another tour, covering 1600 miles in two months, including a stop at the port city of Plymouth, where he made a discovery that would provide abolitionists with another powerful visual aid, one that would become their most famous image: the diagram of slaves packed in the hold of the Liverpool-based slave ship *Brookes*.

The ship is depicted in the diagram with 482 people in the hold where captives appear to be squeezed together as tightly as possible. As bad as it appeared, the image understated the case. To increase their profits, "tight packing" captains of the *Brookes* sometimes oversaw voyages with anywhere from 609 to 740 unfortunates.

Just as today climate scientists sometimes understate predictions of future climate pollution and impacts to avoid criticisms that they are exaggerating, British abolitionists used a conservative figure for the number of captives held in the ship to make the *Brookes* diagram more credible. By April 1787, the image was widely known across Britain, appearing in newspapers, pamphlets, books, and even on posters pasted on the walls of coffee houses and taverns.

Clarkson later commented that the "print seemed to make an instantaneous impression of horror upon all who saw it, and was therefore instrumental, in consequence of the wide circulation given it, in serving the cause of the injured Africans."[12]

A key innovation in activism, such an image had rarely been used as a campaigning tool in this way before. The top-down schematic view of the *Brookes*, with the slaves' bodies squeezed together like sardines in a can, made a huge impact on the British public at the time and continued to circulate for decades to come. Today, the diagram of the *Brookes* is ubiquitous in textbooks, documentaries, and on web pages about the slave trade.

"Part of its brilliance is that it was unanswerable," concludes historian Adam Hochschild. "What could the slave interests do, make a poster of happy slaves celebrating on ship-board? Precise, understated, and eloquent in its starkness, it remains one of the most widely reproduced political graphics of all time."[13]

In terms of high impact, the *Brookes* diagram foreshadowed the "hockey-stick" graph of rising climate pollution developed two centuries later by climatologist Michael Mann. Published in 1998 in a paper that Mann co-wrote for *Nature*, the chart of Northern Hemisphere temperatures going back centuries resembled the shape of a hockey stick. The blade traced warming that rose slowly over centuries at first, but then, beginning in the industrial era, sharply pushed upwards, like the handle of the hockey stick. That trendline of pollution kept going up, with seemingly no end in sight. The image attracted attention from supporters of climate action and science deniers alike. Climate science deniers supported by fossil fuel companies attacked the graph relentlessly, but Mann's research has withstood scrutiny ever since it came out.

Because the image was so simple and so powerful, Al Gore displayed a blown-up version of a hockey stick graph on a large screen in the 2006 documentary *An Inconvenient Truth*. For theatrical effect, Gore stepped onto an electronic lift, which then slowly elevated him up the steeply rising curve of the hockey stick, to show how far "off the charts" carbon dioxide concentrations in the atmosphere had already grown.[14]

Both the hockey-stick graph and the *Brookes* slave ship diagram would make great designs for a T shirt today, but British abolitionists created another image that did make it onto clothing in their own time.

In 1787, Josiah Wedgwood, the pottery manufacturer, asked one of his craftsmen, William Hackwood, to design a seal for the antislavery movement. Hackwood's image depicted an African man kneeling and raising his chained hands in supplication under the tagline "Am I not a man and a brother?" Abolitionists adopted this image as the movement's logo and put it on china, snuffboxes, cufflinks, bracelets, hairpins, medallions, banners, and of course, leaflets, and posters.

Later, antislavery women, who were often excluded from abolitionist groups run by men, came up with their own version depicting a female figure with an appropriately adapted slogan "Am I not a woman and a sister?" Wearing these images became a political statement as much as a fashion statement and helped make the antislavery cause almost trendy.

As popular as these icons were for decades after their debut, they have not aged as well politically as the *Brookes* diagram because many contemporary people feel that images of slaves kneeling in supplication make enslaved people look passive or weak. The enslaved were in fact the first abolitionists, though their role was long ignored in histories of the abolition movement. Enslaved people themselves fought slavery at every step of the way, whether dramatically by mounting slave revolts or through everyday acts of resistance like work slowdowns or breaking equipment.

However art from the early days of British abolition may strike twenty-first century people, the lesson for the climate movement is that such images were emotionally appealing to the public in the late eighteenth and early nineteenth centuries. In a more sentimental age, these images pulled at the heart strings of white Britons. The message that enslaved people of a different race toiling in cane fields thousands of miles across the ocean could be "brothers" of white Britons was an ambitious emotional appeal. But this image helped the appeal succeed, building empathy across the color line between the white public and Black slaves that proved invaluable to the abolition effort.

Every movement needs a good logo, and the climate movement has yet to find an image as memorable as British abolitionists had or even as recognizable as the more famous corporate logos today, from the Nike swoosh to McDonald's golden arches to the apple of a certain tech company.

Perhaps the closest is a famous image of the Earth taken from outer space that appears on everything environmentalist from Earth Day memes on social media to posters, flags, and bumper stickers. Almost as soon as it was shot by Apollo 8 astronaut William Anders in 1968, the environmental movement started trading on the "Earthrise" photo of our planet as seen from the Moon, "the most influential environmental photograph ever taken," according to nature photographer Galen Rowell.[15]

Logos are the simplest form of images that have been key to building recognition for a cause throughout the history of political campaigning. But other images can also become iconic. Propaganda posters like World War I's Uncle Sam pointing at the viewer and announcing, "I Want You for U.S. Army," and World War II's Rosie the Riveter helped recruit men and women into all-out war efforts. By contrast, horrifying photographs helped turn viewers against war, photos like Matthew Brady's view of the battlefield dead after the Battle of Antietam in 1862 during the Civil War and Associated Press photographer Nick Ut's shot of terrified children running from the site of a Vietnam napalm attack in 1972.

After years of relying heavily on scientific charts and graphs, the climate movement has started using images of people to make the problem of invisible greenhouse pollution in the atmosphere clearly into a problem for human beings. A powerful example is the winner of the One-Shot Climate Change Competition in 2020 by Indian photographer Daksh Sharma. It's a Black-and-white shot of a little boy holding a handkerchief over his nose and mouth standing in front of the iconic Gateway of India in Mumbai on a smoggy day. "When the air becomes heavy with carbon and other pollutants and your lungs sear with the pungency of the air which was essence of life till yesterday," comments Sharma. "Now it appears hell-bent on taking it away...It's just a breath!"[16]

Another way that British abolitionists found to preach beyond the choir and reach a larger public audience was through popular music. In an age before copyright restrictions, it was common to set new lyrics to well-known tunes, making the new songs easy for many people to sing. An example from abolitionists was "The Negro's Complaint," a poem by William Cowper set to the tune of a popular song called "Admiral Hosier's Ghost":

Forc'd from home, and all its pleasures,
Afric's coast I left forlorn;
To increase a stranger's treasures,
O'er the raging billows borne,
Men from England bought and sold me,
Paid my price in paltry gold;
But, though theirs they have enroll'd me,
Minds are never to be sold.

Song is another art form that has been used effectively by activist movements, political parties, and government propagandists alike. Woody Guthrie's "This Land is Your Land" may be the closest thing to an anthem for the environmental movement. Like many activist causes on the left, environmentalism has drawn heavily for its music from mid-twentieth century folk ballads, spawning generations of soulful tunes on acoustic guitar. Seeking to speak to new generations, musicians have recently started producing songs about the climate crisis in genres from hip hop to rock to electro pop. Perhaps reflecting the more fragmented music market of the internet age, the climate movement has yet to find an anthem as powerful "This Land is Your Land" that's familiar to a wide enough audience and easy enough for anybody to sing.

British abolitionists of the eighteenth and nineteenth centuries produced songs and other pop culture that had the widest appeal, helping to bring new people into their movement. To recruit new members outside of the usual suspects, the climate movement would do well to follow their example. We need climate music that goes beyond artistically sophisticated performances meant to be watched on web videos and spills out into simple, old-fashioned songs that people can sing along with. More about that in the next chapter.

Using songs, poetry, visual art, and a catchy logo helped British abolitionists make an emotional appeal to complement the pamphlets, lectures, and public meetings that were traditional in a political advocacy campaign. Enlisting popular culture in the service of a cause was a key innovation of British abolitionists and it paid off in terms of public support. By the end of 1788, 183 petitions had been sent to Parliament, signed by tens of thousands of people.

Even port cities that hosted slave ships or inland cities with factories that relied on cheap supplies of sugar, tobacco, indigo, and cotton grown on slave plantations in the West Indies sent out antislavery petitions packed with signatures. Nearly 20 percent of the population of cotton-mill town Manchester—more than 10,000 citizens—signed a petition calling for an end to the slave trade. That gave abolitionists the popular base of support to turn public outrage into political change.

As a result of lobbying by abolitionists, Parliament held hearings, debated legislation, and ultimately began to pass antislavery bills. Abolitionists knew that they wouldn't get everything they wanted right away, so they took a gradual approach as we've mentioned. In 1788, just a year after abolitionists published the diagram of captives stuffed into the hold of the *Brookes* slave ship, Parliament passed a bill passed to limit the number of slaves that a ship could carry, though the regulations were never well enforced.

As the tide of public opinion began to turn against slavery, elite opinion followed suit. A lobbyist for the West India slaveowners even reported back to his clients in Jamaica that "the Press teems with pamphlets upon the subject...the stream of popularity runs against us."[17]

The Very Model of a Dirty Industry Lobby

The following year, in 1789, the abolitionists' champion in Parliament, William Wilberforce, was finally ready to submit his bill to abolish the slave trade. Demands from the planters' lobby for more data—a popular political delaying tactic then as it is now—caused hearings in the House of Commons to drag on for two years. This gave the planters and their allies among merchants and shipping interests time to organize their own propaganda and lobbying campaign to defend the slave trade and slave agriculture in Britain's Caribbean sugar colonies.

Liverpool alone, a port city whose economy was heavily reliant on slavery, spent £10,000 on lobbying to protect the slave trade, equal to about $1.4 million today. Recognizing that they would have to spend more money on lobbyists and propaganda to counter the threat from abolitionists, the West India Committee raised the levy on its member planters from a penny to sixpence per barrel of sugar imported.[18]

To justify their profitable industry, slaveowners and their allies had to make it seem that slavery was not so bad and might even be good. Or at least that slavery was a necessary evil for white people to enjoy a prosperous economy. For example, in Britain, a proslavery candidate for Parliament turned this argument into a campaign jingle:

If our slave trade had gone, there's an end to our lives,
Beggars we all must be, our children and wives,
No ships from our ports, their proud sails e'er would spread,
And our streets grown with grass where the cows might be fed.[19]

How little things change. In the global industrial economy, steel container ships have replaced wooden sailing ships, but the sentiment remains the same. Big businesses complain that if government, representing the will of the public, forces those businesses to stop doing nasty things that the economy will suffer. But usually when a dirty or harmful industry has to reform, it turns out that only those businesses themselves suffer and everybody else is fine. Most people may even be better off. It was true when slavery was abolished, and it's also proven true when fossil fuel companies have had to stop polluting.

When President Biden moved to ban new oil and gas drilling on federal lands, a small portion of all the places where companies can drill inside the United States, the American Petroleum Institute claimed the move would "risk hundreds of thousands of jobs and billions in government revenue for education and conservation programs."

But the oil lobby is as good at exaggerating today as the West Indian planters' lobby was. It turns out that the API was inflating job loss numbers. In 2020, oil and gas companies employed about 160,000 across the United States, but they claimed that they'd lose tens of thousands more jobs than that in just a few states. "API is claiming that more people would lose their jobs than the industry actually employs. Even accounting for ripple effects on related industries, it is a staggering claim," documented Nick Cunningham in *DeSmog*.[20]

As we'll see later, the loss of first the slave trade and then of slave agriculture itself didn't destroy the British economy. On the contrary, ending slavery provided a huge economic stimulus.

But the slave lobby had other arguments for why they should be allowed to keep buying and selling human beings. One proslavery witness testified in Parliament that Africans were so glad to be escaping their continent's barbarism that "Nine out of Ten rejoice at falling into our Hands." In contrast to former slave Olaudah Equiano's picture of time on a slave ship as a voyage of the damned where the face of every captive showed "dejection and sorrow," former ship captain James Penny made the infamous Middle Passage from Africa to the Americas sound like a luxury cruise.

Again, industry propaganda was a little too enthusiastic and ended up crossing unintentionally into self-parody:

If the Weather is sultry, and there appears the least Perspiration upon their Skins, when they come upon Deck, there are Two Men attending with Cloths to rub them perfectly dry, and another to give them a little Cordial...They are then supplied with Pipes and Tobacco...they are amused with Instruments of

Music peculiar to their own country…and when tired of Music and Dancing, they then go to Games of Chance.[21]

The narratives of Equiano and other former slaves made it easy for abolitionists to refute such absurd claims from the planters' lobby. Nonetheless, it still took abolitionists years to build up enough support in Parliament to get Britain out of the slave trade. Wilberforce's first attempt to pass a bill to abolish the trade in 1791 failed on a vote of 88 to 163. During the next two decades, Wilberforce would introduce nearly a dozen more bills to abolish the slave trade, a time when Britain was pulled into momentous world events including the French Revolution and Napoleonic Wars, which both helped and hurt the abolitionists.

Making World Events Work to their Advantage

We've seen how, trying to prevent rebellious slaves in Haiti from declaring the colony independent, the French National Convention issued its Decree of 6 Pluviôse (4 February) 1794 abolishing slavery in its colonies including Haiti. After that, abolition became associated in Britain with the French revolutionaries against whom Britain had recently declared war.[22]

This was a boon to West Indian slaveowners, who were now able to claim that abolition was a dangerous doctrine connected to Britain's wartime enemy. Many in government agreed, putting pressure on abolitionists to stand down, which they did for half a dozen years during the war with France.

In the meantime, the worst fears of West Indian planters were realized. The Haitian revolution did inspire slave revolts across the Caribbean and in some British colonies there, to which Britain had to respond. Altogether, between 1793 and 1801, Britain sent more 90,000 soldiers to try to suppress restive slaves in the Caribbean, a larger force than it had sent to put down rebellion in its North American colonies in the 1770s (what we call the American Revolution), just a couple decades before.

In the Caribbean, British forces fighting rebellious slaves experienced heavy losses to death, injuries, disease, and desertion. The difficulty of suppressing slave revolts in the West Indies sent a powerful message to people back home that the costs for maintaining the slave trade were too high. "Many in Britain began to question whether more lives and resources (substantial public funds were being used to pay for soldiers, forts, and naval bases) should be spent sustaining the trade."[23]

This message was further reinforced after the British withdrew from Haiti. A few years later, the French under Napoleon foolishly tried to retake their former colony and revoke the abolition of slavery there. The French lost 50,000 troops in Haiti, more than would die at Napoleon's famous defeat at Waterloo. It became clear to abolitionists and the broader public alike that the wealth and military might of neither Britain nor France were sufficient to keep them from being defeated by an army of self-emancipated slaves.

Crises get the public's attention. Just as a major climate disaster like Hurricane Katrina or California wildfires has given new urgency to climate solutions in recent years, so slave revolts in the Caribbean gave new urgency to the antislavery campaign in the early nineteenth century. Following the incremental strategy outlined at the birth of the movement, first the abolitionists lobbied not to end slavery, but just the Atlantic slave trade. And within that trade, abolitionists pushed an incremental approach that was more politically realistic as well.

Taking a patriotic angle, abolitionists asked not to end the slave trade entirely, but only the portion of the slave trade with France and its allies. Since Britain was still at war with France, it was hard for the planters to object to a measure that would deny resources to the enemy. The Foreign Slave Trade Bill passed into law in 1806. This set the stage for Wilberforce to resubmit his bill to abolish the complete British slave trade, which finally passed the following year and went into effect on January 1, 1808.

Inspired by legislation in the United States to end its own participation in the international slave trade on the exact same date, January 1 became a holiday celebrated for decades to come by antislavery activists on both sides of the Atlantic. But neither nation had abolished slavery itself yet. For many abolitionists, the end of slavery was in sight. Yet not all signs were encouraging.

In Britain's colonies in the West Indies, the slave population continued to grow. British planters, deprived of a source of new slaves, improved conditions to take better care of the slaves they already had and to help their families increase in size. Even though such reforms made slavery less brutal, hundreds of thousands of people would still be treated like property and would still be subjected to a cruel regime of forced labor in the grueling work of sugar production.

For abolitionists, ending slavery altogether would come next. Unfortunately, their victory against the slave trade combined with complications from the endgame of the Napoleonic wars sidelined abolition for another decade and a half. After the slave trade was ended, many Britons thought that slavery would die out on its own, just as many people today think that fossil fuels will just fade away from market forces because clean energy will become ever more affordable and practical. But government action will be needed to end fossil fuels in the future just as legislation was needed to end slavery in the West Indies in the nineteenth century. Abolitionists knew this, but they also knew that just after abolishing the slave trade, the political environment wasn't ready for complete emancipation yet.

As interest in fighting slavery dissipated among the British public, some activists switched over to trying to mitigate the problems of slavery, pushing for government oversight to improve the treatment of the enslaved and limit their working hours. The Crown agreed to create a new government position, His Majesty's Protector of Slaves, to investigate reports of abuse by planters and overseers and enforce new regulations for humane treatment of the enslaved. Obviously, such an approach had its limitations. Trying to make bound labor less oppressive risked fooling the public into thinking that slavery could be reformed into a benign institution and concluding that there was no hurry to abolish it.

It's like people today who know that fossil fuels are destroying the climate but aren't willing to push for full abolition of oil, gas, and coal. Instead, they support such ways to allegedly burn fossil fuels safely like capturing carbon pollution and sequestering it in old oil wells or coal mines. Unfortunately, as we've seen, scientists have determined that most plans for carbon sequestration are impractical. Capturing and storing carbon dioxide gas is unsafe (what's pumped down will eventually come back up) and it's more expensive than just replacing fossil fuels with clean energy.

The climate movement has waxed and waned over its history of three or four decades. In 2007, the Sierra Club even partnered with natural gas drillers to promote gas as a cleaner fossil fuel, a "bridge" to ease the transition from dirty coal to clean renewables. As part of its Beyond Coal campaign to shut down coal-fired power plants, the nation's largest environmental group took more than $26 million from people and companies connected to Chesapeake Energy, one of the largest gas producers in the country, before reversing course in 2010. Later, the Sierra Club came to regret its support for natural gas.[24]

But people in frontline communities never stopped fighting to stop oil refineries, plastics plants, and fracking operations from polluting their air, water, and land. In the same way, even while white Britons lost steam on antislavery, in the West Indies, slaves themselves didn't stop fighting for their own freedom.

Slave revolts in Barbados in 1816 and in Guyana in 1823 led to the destruction of millions of pounds in crops, buildings, and equipment along with the deaths of hundreds of slaves. Angry planters in Guyana pinned the blame for rousing slaves on a young white missionary, John Smith, who was sympathetic to the plight of the enslaved in the colony. Sentenced to be hanged, Smith died in prison before a pardon came through from King George IV, making the missionary a martyr for British abolitionists. They were able to use the death of a fellow white person to arouse more sympathy among the British public to the antislavery cause.

Though the antislavery movement lost its way after the slave trade was abolished, dedicated activists kept circulating petitions and lobbying the government to pressure foreign powers like France and Spain to follow Britain's example and get out of the slave trade as well.

But it wasn't until 1823 that scattered groups of gradual and immediate abolitionists got together to form a new umbrella organization, the London Society for Mitigating and Gradually Abolishing the State of Slavery Throughout the British Dominions. This un-catchy title expressed the group's incrementalist approach, helping to present what was a radical movement as conservative, reasonable, and non-threatening to the powers that be—especially the economic powers. After all, many government leaders, members of Parliament, and middle-class families were invested in West Indian plantations. And under Britain's aristocratic system of limited democracy at the time, where only men with large amounts of property were allowed to vote, earning support from the well-off was crucial.

Fortunately for abolitionists, changes in Britain's voting system that enfranchised the middle class soon came along that made it less important to cater to people with the biggest economic stake in slavery. After years of activism by workers, Parliament passed the Reform Act, which expanded the electorate to the middle class. Though it did not include women, the Act did enfranchise a larger group of men, many of whom had no direct investment in the West Indian sugar industry.

As a result, the elections of 1832 brought new antislavery members to Parliament while half of the members who owned plantations lost their seats. Imagine today, in the United States, if Congress passed significant campaign finance reform to nullify the *Citizens United* case that allowed Big Oil to spend unlimited amounts to buy political candidates. If ordinary Americans had the power they're supposed to have in a democracy, judging by opinion polls, serious climate solutions would quickly become the law of the land.

In Britain in the 1830s, the stage was set for one more national disaster to push abolition over the finish line in Parliament. And that meant one final bloody slave rebellion in the Caribbean.

Just after Christmas Day in 1831 in Jamaica, Britain's richest West Indian colony, 31-year-old Samuel Sharpe, the enslaved deacon of a Baptist mission in Montego Bay, led a rebellion of 20,000 slaves that took over large parts of the island. After nearly a month of fighting that killed 200 slaves and 14 white people while destroying £1.1 million in property on 200 plantations, British troops regained control of the island. Afterwards, they executed 312 rebels, including the leader of the rebellion, Sharpe, who said, before he died, "I would rather die upon yonder gallows than live in slavery."

Slaveowners won the battle against Sharpe and his slave army in Jamaica. But it became clear that slavery's days were numbered in the British Empire. The insurrection in Jamaica warned the British government that it might soon have to wage an all-out war to prevent the loss of one of its most important colonies, just as the French had lost Haiti less than 30 years earlier. "The present state of things cannot go on much longer...every hour that it does so, is full of the most appalling danger...Emancipation alone will effectively avert the danger...in the meantime it is but too possible that the simultaneous murder of the whites upon every estate in the island...may take place," warned a top official at the British Colonial Office.[25]

This was an opening for abolitionists to achieve the final victory they had fought so hard to win over half a century or more. Only two weeks after the death of William Wilberforce, who had spent his long political career trying to pass antislavery legislation, Parliament finally passed, and the King gave his assent to, the Slavery Abolition Act of 1833. One of the most historic pieces of legislation in human history, the Act took effect on August 1, 1834, ensuring that 800,000 people could no longer be bought, sold, or owned throughout most of the British Empire.

Ending slavery was a huge victory for enslaved people and abolitionists alike. After centuries of violence in slave rebellions along with the everyday violence of slaveowners to keep their workers in chains, abolition was accomplished peacefully. That's because unlike in the United States, slaves in the British Empire were emancipated with the concurrence, no matter how grudging, of slaveowners, rather than at the point of a bayonet as in the U.S. Civil War. Offering the West Indian planters cash compensation along with a gradual schedule for emancipating slaves helped neutralize opposition from the proslavery lobby.

Paying Owners to Walk Away

The Slavery Abolition Act had two clauses which some abolitionists opposed vehemently. Yet these compromises with the West Indian planters' lobby were almost certainly crucial to making British abolition happen when it did instead of years or decades later.

First, the Act awarded cash payments of about £20 million to slaveowners in compensation for the loss of their property in enslaved people, an amount equal to as much as $414 billion today.[26] Cash payments represented 45 percent of the value of each slave, determined in what was perhaps the first instance of detailed statistical analysis by the U.K. government, based on a listing of every slave by age, sex, skill level, health, and level of productivity on their plantation. Meanwhile, slaves themselves got nothing.

Second, slaveowners would get another form of compensation in the form of additional years of work from their former slaves. This was to be the slave's own contribution to his or her freedom, which represented another 35 percent of each person's assessed value. Under the terms of the Slavery Abolition Act, on January 1, 1834, slaves wouldn't become free right away. They would temporarily become "apprentices," most of whom would be forced to keep working for their old masters without pay until 1840.

Black workers, who hated this requirement, mounted widespread strikes in the West Indies. Supporting their protest, back in Britain abolitionists got more than half a million signatures. Again, the activist campaign was successful. Under pressure, Parliament ended the apprenticeship period two years early and declared all former slaves forever free on August 1, 1838.

Together, cash payments from the government and years of free work from each formerly enslaved apprentice covered 80 percent of the assessed value of each slave, leaving the owners to take a 20 percent loss. Morally, such accounting may sound absurd if not obscene to modern ears. It is not possible to put a monetary value on a human being. Yet, it was exactly this kind of accounting that convinced British Caribbean slaveowners to finally walk away from their long fight against abolition.

Recently, a similar debate has occurred in the fight over climate solutions. While economists have tried to put a dollar value on life support "services" provided by nature—clean water filtered by plant life or oxygen from trees, for example—some environmentalists have objected that it's immoral to put a price on nature, whose benefits should be free to all. But economists and even some scientists say that unless we do show that trees, animals, rivers, and mountains offer benefits to people that are measurable in dollars and cents, then the public will have a hard time understanding why they should care about protecting such features. People might say that nature is priceless, but all too often they act as if it's worthless.[27]

Back in 1834, compensating 46,000 slaveowners was hugely expensive, a sum representing 40 percent of that year's national budget. The Bank of England financed compensation payments with bonds for £15 million, a debt that was only fully paid back in 2015. As we've seen, retiring that debt reignited the long running controversy over giving payments to slaveowners instead of freed slaves.[28]

University College London set up the Legacies of British Slave-ownership database, allowing the public to search records of people who received compensation payments in the nineteenth century and connect them to their descendants living today. Among those revealed to have benefited from slavery are ancestors of former Prime Minister David Cameron, novelists Graham Greene and George Orwell, and poet Elizabeth Barrett Browning. George Orwell's great-grandfather, Charles Blair, received £4,442, equal to $4.12 million today, for the 218 slaves he owned.[29]

Pro-climate investment banker Richard Barker has concluded that payments to slaveowners injected powerful stimulus into the growing industrial economy. "The compensation payment was put into the investing classes and was a large fiscal event." Compensation money went into factories and railroads, which "helped turbocharge Britain's industrial development and wealth way above that of its competitors."[30]

Minus any focus on equity, this is the same idea behind today's Green New Deal—that spending money on clean energy or cleaning up pollution will also stimulate the economy, creating millions of green jobs for workers, even helping some from fossil fuel industries transition into the new clean economy.

Unlike the plans for the Green New Deal, though, none of the slavery abolition stimulus went to the workers, the freed Black people who toiled without pay, as so many of their ancestors had, in the sugar cane fields of West Indian plantations. Since former slaves received no compensation in either money or land, some had little choice but to earn a living working for their former masters at low wages. Island governments levied new expenses including taxes and planters charged rent on huts.

To escape the system of low wages and new costs, many former slaves retreated into the forested area long neglected on many islands, cleared the land, started homesteads, and fed themselves from their gardens, basically living off the grid. Wherever they lived, many freed people found room to maneuver in Caribbean economies that allowed them to acquire land, start businesses, and even emigrate to other nations to seek jobs.

Compensation vs. Reparations

British abolition through compensation in the nineteenth century teaches two important lessons for climate solutions today.

First, it's both morally wrong and economically unsound to compensate only owners but not provide reparations to the people who suffered under the old, unjust system—former slaves back then, workers or communities on the front lines of fossil fuel pollution and climate impacts today. Freedom to the enslaved would have meant much more if freed people in the Caribbean had been given land and money to buy tools, equipment, seed, and other necessities to start their own farms. And Caribbean economies, long held hostage to the needs of the descendants of the same class of white landowners who prospered under slavery, would be more prosperous today if wealth had been more evenly distributed in the colonial past.

Second, as morally repugnant as it seemed to abolitionists to pay slave masters compensation for their lost "property," this was the key component to making British abolition happen when it did, instead of decades later. Compensation to slaveowners allowed the government to free slaves while respecting the principle of private property, a bedrock of the British economy back than as much as it is of the world economy today. And though it took two centuries of bloody slave rebellions and everyday violence suffered by enslaved people, at the end, the British Empire ended slavery without anything like the carnage of the American Civil War. This saved much human suffering, and it saved a lot of money. "The cost of damage to property and lost productivity in the Civil War turns out to be between 12 and 18 times more expensive than the compensation that was agreed in Britain," says Barker.[31]

He contends that without compensation, the West India lobby would have continued to fight against abolition, which probably would have kept Black workers in bondage for decades longer. Without compensation to owners, support for slavery also would have extended into British society far beyond the planters in the islands, since one in five wealthy people back home in Britain had investments in Caribbean plantations.

Today, just think of all the people who own investments that would lose value if fossil fuel use declined rapidly or even ended abruptly— stocks in oil majors like Exxon but also in chemical and plastics makers that would need to scramble to find affordable "feedstocks" to replace petroleum products. And what about small businesses like gas stations?

It's not just cigar-chomping fat cats who can send out lobbyists to oppose climate solutions that might cut into profits of fossil fuel interests. It's lots of ordinary business owners and workers who can call their member of Congress to complain that hurting the oil industry will deprive millions of people indirectly connected to fossil fuels of a way to make a living.

The analogy between British slavery abolition in the 1830s and climate solutions today offers both good news and bad news according to investment banker Richard Barker.

To start with the bad news, international agreements signed at climate conferences from Stockholm in 1972 to Rio de Janeiro in 1992 to Paris in 2015 were not ambitious enough to deal with the scale of the threat to the climate. Due to political pressure from oil producers, these agreements were able to go no further than trying to make climate change less bad by regulating down the emissions of greenhouse gases.

"We have been constrained and manipulated by the oil lobby, the coal lobby, the international shipping industry," Barker says. Facing political opposition from a powerful industry, on climate "we haven't necessarily moved beyond what was the abolition in the trade in slaves. What we need to do is look forward to the complete equivalent of the abolition of slavery."[32]

Now we are at the point in history where we must not just restrict oil but abolish it completely. Just as slave rebellions helped convince Britons in the 1830s that slavery was no longer an institution they could support, so recent climate disasters have convinced more people that fossil fuels must go.

"Mother nature is rebelling from all of the CO_2 we're pumping into the atmosphere. We're seeing increasing numbers of significant events of losses worldwide related to storms, fires, etc." The human cost of climate chaos is too high, and for Barker, the economic cost is too high also.

> The governor of the Bank of England has identified that climate change can pose a significant risk to global financial security...perhaps like the ferocious slave rebellion in Jamaica finally forcing people over the line, perhaps what we'll see is a major weather-related catastrophe. A major climate-induced financial shock in the markets will push governments and financial markets to make the final move to the equivalent of abolishing slavery.[33]

The partial compensation that the British government paid to slaveowners can serve as a precedent to induce fossil fuel companies and their allies to agree to leave most of the remaining oil, gas, and coal in the ground. "We need to price carbon properly, we need to tax carbon effectively, the polluters need to take their loss and finally, though unpalatable for some is that we probably need to pay compensation," Barker says.

What form would that compensation take?

"There's a compensation number, a period of time the slaves are going to work for you, and an absolute loss," Barker told me, using the analogy of compensation to British slaveowners in 1833 to what oil companies might expect soon.[34]

He envisions compensation for fossil fuel companies to abolish oil not in terms of cash payments like those given to British slaveholders. Instead, compensation today would come as additional time for oil, gas, and coal companies to operate before being shut completely down, more like the unpaid work that recently freed people in the West Indies had to offer their masters under the transition period of apprenticeship.

Barker went on to tell me how fossil fuel companies may choose compensated abolition as the lesser of two evils from their point of view—either plan to gradually down voluntarily with the chance to recoup some of your loss or try to fight but risk taking a much bigger loss later.

> Today, an oil company is valued based on its current cash flows and reserves. You're running a huge risk as a corporate leader that carbon will be taxed in the near future or that you might be prohibited from exploring for or extracting oil. Companies need to understand that there's an absolute risk of discontinuity in their business today.

> Do you want to take that risk? Or should we insulate you from that risk? You could align your operations with the Paris Accord. You'll wind down exploration and we'll put in a tax structure that will encourage this. If society provides some help to phase out their operations, then perhaps dirty energy companies can be convinced to absorb the remainder of the value of their future profits as a loss. [35]

Hard Fighting before the Armistice

The day for negotiation about that kind of compensation may come soon. But for now, dirty energy companies haven't suffered enough yet. They're not ready to talk about shutting down their industry. Instead, they still think they can find ways to do business for decades to come by turning oil into ever more plastic products or burning gas to produce hydrogen or convincing the public that they can keep pumping oil but start capturing its pollution and safely store it underground.

Despite hype created by their well-funded propaganda machine, most or all the ideas that would let the industry continue to produce fossil fuels supposedly safely for the climate are dead ends— technologies that are unproven, or too expensive, or just plain don't work. But if they can convince the public and elected officials that these ideas have promise, the Oil Power can continue to fry the climate. We must not be fooled. We need to shut down fossil fuels altogether. There's no other way to save the climate according to scientists.

To bring oil, gas, and coal companies to the table to negotiate outright abolition of fossil fuels, people who care about the climate still need to campaign hard to weaken the political power of the industry.

Under pressure from the public, government and industry alike have been switching to clean energy, pulling business away from fossil fuel companies. "Those companies are already seeing their profits decline," Barker told me. "As they become weaker and weaker financially, they'll also become weaker politically. At some point, Exxon and their friends will be willing to come to the table for complete oil abolition rather than suffer a worse alternative—complete loss without any recompense."[36]

The Biden administration and Congress made a good start in early 2021 with executive orders and legislation to reduce or end fossil fuel subsidies. Though the subsidies are worth billions of dollars, cutting taxpayer support for dirty energy may be less about taking money away from fossil fuel companies than about influencing public opinion.

"The dollars at stake here are not going to hurt the industry," according to Andrew Logan, an expert on oil and gas at Ceres. "What they really don't want to do is become the next tobacco."

If Biden's executive order and legislation from Congress killing oil subsidies sends the message that the federal government has switched from pro- to anti-fossil fuels, that message could ripple across the economy, encouraging banks and investors to pull away from oil and gas companies in the future, just as legislation to ban cigarette ads on TV and radio in 1970 started a long period of decline for American cigarette marketing.[37]

Fossil fuel companies will only feel the pressure for full abolition if those companies suffer financially while losing the support of the public, just as the Caribbean planters did back in the 1830s. To hit dirty energy companies where it will hurt most and weaken them both financially and politically enough to bring them to the table sooner rather than later for complete abolition, Barker wants government and the financial sector alike to move funding away from fossil fuels and into clean energy.

To transition from a dirty to a clean economy, the nations of the world need to invest in the post-abolition economy to the tune of $90 trillion according to Barker. That must involve compensation to frontline communities and workers, the missing component of British slavery abolition. Oil abolition can build back better with an approach focused on climate justice like the Green New Deal.

Finally, any deals that allow fossil fuel companies to keep operating through market-based programs to cut or offset pollution must be scrutinized. Programs must work to reduce carbon rather than just to boost the PR of dirty energy companies. According to Mark Jacobson of Stanford University, simply using fewer fossil fuels is more effective to cut pollution than most market-based climate solutions. "Carbon taxes being pushed by the oil and gas industry, just like their push for carbon capture—it's just useless scams. They are just not anything that will actually help solve the problems."[38]

Leading scientists and climate experts including former NASA climatologist James Hansen support putting a price on carbon if it's designed properly to cut pollution rather than just move it around. But oil companies' apparent support for a carbon tax has been dishonest. In the summer of 2021, Greenpeace released a video that contained an undercover interview with an Exxon lobbyist named Keith McCoy who claimed that the company didn't really want a carbon tax but that they only claimed to support it in public for appearances because they thought it would never be implemented.[39]

Only by abolishing oil while investing in the new technologies of a clean economy can the nations of the world avoid a global catastrophe worse than the U.S. Civil War. If we succeed, and do it soon enough, then humanity can enjoy a century or more of prosperity.

To succeed, oil abolitionists will need to follow the example of slavery abolitionists in both Britain and the United States and find ways to build wider public support. That's why the next chapter will talk about the leaders and the strategies that brought victory to American abolitionists in the nineteenth century and will help oil abolitionists win in this century.

7

Heroes of Abolition
Then and Now

We forget that it was the religion of abolitionists that led them not to
pray against slavery but to stand against slavery.

—Rev. William J. Barber II

T he fight against climate breakdown began with science and
scientists. For those in the know, their names have been
enshrined on a roll of honor: Joseph Fourier, John Tyndall,
Svante Arrhenius, Guy Stewart Callendar, Roger Revelle, Charles
Keeling, and James Hansen.[1] Ever since the most recent of those
climatological giants, James Hansen, testified in Congress in 1988 that
climate heating was increasing, that it was dangerous, and that human
activity was the main cause, the movement to save the climate has been
about spreading awareness of the facts and urging society to act on
those facts. Climatology, oceanography, and engineering provided data
and tools to achieve an eminently rational goal—to cut greenhouse
pollution enough to stabilize the climate and to do it fast enough to
avoid catastrophe.

Much progress has been made. The world will long owe a debt to
the scientists who sought to uncover the laws of the climate system and
warn industrial society of the penalties for breaking those laws. Now it's
time for those of us who join those scientists in wanting to save
humanity to pitch in. We need to build on their work and do our own
work to mobilize citizens across America to demand serious action on
climate solutions from federal, state, and local governments. We must

also stand with those brave and committed scientists against the lies and political pressure of Big Oil.

People who want to save the climate must become as good at talking about the truth as those companies that would destroy the climate are at hiding the truth. Fossil fuel companies have done well at reaching ordinary Americans by appealing less to the head and more to the heart. That's why they claim that Coal Keeps the Lights On,[2] that petrochemicals Feed the World,[3] that Plastics Make It Possible,[4] and that there's a Moral Case for Fossil Fuels.[5] To climate activists, such slogans are clearly obscene. But such messages can confuse citizens who are not experts on the issue.

Oil, gas, and coal companies say that they save lives and make modern life comfortable and convenient, while offering the world's poor hope for a brighter future. Their ads show kids running in fields of flowers, clean-cut guys working as engineers, astronauts blasting off in rockets, artificial hearts pumping, and athletes running on tracks.[6]

The climate movement has been criticized for talking too much about science. A few key scientific facts were always necessary to explain the greenhouse effect and help the public understand the depth of the problem, its cause, and the solutions available now. Credible science has been even more necessary on climate than on less scientific and engineering problems—say, making air travel safe or protecting the internet from hackers—because the fossil fuel industry didn't cooperate but insisted on questioning the facts and spreading lies.

Yet, it's a fair criticism to say that campaigning from climate advocates has included a lot of intellectually challenging content, not just about climate science but also about public policy and energy technology. As a result, the average person who prioritizes climate above other political issues is likely to be better educated than most Americans.

Things haven't changed that much since the nineteenth century. Abolitionists were brainy, geeky, well-read types who valued the truth as expressed by experts qualified to talk about the most respected disciplines of their day. Back then, that usually meant religion but also started to include science. But antislavery activists quickly learned that they couldn't defeat the Slave Power with persuasion alone.

They needed to exert political and economic pressure to change laws against powerful special interests who didn't want those laws changed. Abolitionists had to evolve from "moral suasion" to confrontation. Ultimately, in the U.S. if not in Britain, abolishing slavery took war. To build the pressure to go up against powerful opponents and their big money, abolitionists knew they needed to recruit a much bigger group of supporters from the wider public. And that would require an appeal less to the head and more to the heart.

Sometimes the climate movement seems to pride itself on being leaderless, a people-powered movement of a billion points of light on the internet. Instead of names of a few recognized and experienced leaders, pro-climate media and groups put out names of dozens of "emerging leaders."

Or maybe they put out no names of people at all, but just names of organizations. "The Stop the Money Pipeline coalition is over 150 organizations strong," explains the website of a "movement" to pressure banks to stop funding fossil fuel projects. A very worthy goal, but who's behind it? You'll look in vain to find the names of any humans, but you can see 27 rows of logos from environmental lobbying groups probably financed by philanthropic foundations but with few or no ordinary people as members. That's why you've probably never heard of most of these groups. It's insider stuff.[7]

The no-leader approach sounds democratic and egalitarian and very much in line with the kind of anti-hierachical philosophy of activists on the left since the 1960s. But it's boring and confusing for people outside of the core of leaders of climate lobbying organizations and their funders in philanthropic foundations.

Ordinary people remember names, faces, and stories of other people, especially extraordinary leaders. Imagine the civil rights movement without Martin Luther King or Malcolm X or John Lewis. Yes, there were hundreds of other leaders and millions of other activists involved. Everybody deserves some credit. It was the same with all successful movements from the past. Yet, members of those movements still knew there was immense value in highlighting leaders.

Abolition leaders shone a spotlight on the evils of slavery and the glories of freedom. But those same leaders were also the subject of abolitionist storytelling, each leader serving as the example of what an engaged citizen should look and sound like. The antislavery movement took every opportunity to celebrate its heroes, both Black and white. Highlighting leaders of conscience was a powerful way to inspire supporters to put in the hard work, year after year, over the decades that it took to win abolition on both sides of the Atlantic.

In the United States, "Eminent Opponents of the Slave Power" was a print produced in 1864 during the Civil War. It portrayed a dozen white abolitionists including former President John Quincy Adams, antislavery preacher Henry Ward Beecher, editor and publisher William Lloyd Garrison, and poet John Greenleaf Whittier.

Despite the old-timey phrasing of its title, this image offers a powerful example today of the value of celebrating a movement's leaders. The title is not a modest one. Instead, it's bold and creates narrative tension by setting white male abolition leaders up as enemies of the "Slave Power," an evil and menacing-sounding force that didn't just enslave Black Americans but also hacked at the very roots of American democracy and freedom for white people. The white abolitionists in the print weren't just fighting the nameless system of slavery. They were fighting The Slave Power, a menacing enemy that had to be named to conjure up supporters' outrage.

This image was joined by others highlighting heroes that circulated among abolitionists. Printed after the Civil War, "Distinguished Colored Men" depicts activists for abolition like Frederick Douglass along with public officials and diplomats appointed or elected during the Reconstruction period after the war.

Meanwhile, Sojourner Truth raised money for her lecture tours by selling visiting cards with her photo and the haunting inscription "I sell the Shadow to Support the Substance."[8]

Fictional heroes got their due too. Prints depicting Uncle Tom, Eliza, Eva, Topsy, and other characters from *Uncle Tom's Cabin*, the blockbuster bestselling novel of its day, circulated beyond abolition circles outward across the reading public in the Northern states and around the world.

And engravings of heroes of abolition cut from the pages of newspapers became wall art, whether depicting John Brown's raid on Harpers Ferry just before the war or, after Fort Sumter, Black soldiers in units of U.S. Colored Troops fighting in battle.

Such images encouraged both admiration and emulation. Examples of abolition leaders inspired ordinary people to do extraordinary things. Today, people who care about climate need inspiring leadership as much as abolitionists did back then. Now, the United States and the rest of the industrialized economies of the world must cut their emissions in half by 2030, a mission that will require fundamental transformations in energy, the economy, construction, finance, and more.

That's an almost impossibly ambitious goal for such a tight deadline. Experts say we can do it, but it's going to take a nearly superhuman effort. Then, even if we're successful in the short term, there will still be plenty of work to save the climate for decades afterwards. The fossil fuel industry has proven through decades of booms and busts that it is resilient. Dirty energy producers have endured and then successfully fought off intense public criticism ever since they started mining and drilling. The climate movement must become much more powerful in the coming years.

Even so, oil, gas, and coal companies will probably still be doing business in some form in 2030, with hopes to keep going as long as they can. So, saving the climate is both a short game and a long game. Igniting a burst of public support now and then keeping that support burning for another half century will be key to winning both games.

In the short term, the public needs to be mobilized for a wartime level of transformation of the economy. Changes will bring many opportunities in green jobs and green stimulus that will make many Americans more prosperous. But fossil fuel companies will continue to fight, even if only through deceptive attempts to convince us that they're "part of the solution," when they're just finding clever new ways to pollute. In the longer term, a broad consensus must be maintained to stay the course through ups and downs of the economy resisting many temptations to relent.

Archetypes that Helped Abolition Win

The concept of archetypes has entered everyday life and only requires a little background to understand. Etymologically, the word archetype is a compound of two Greek words: *arche* is an origin which cannot be seen directly and *tupos* is a visible impression or manifestation of that origin.

Psychologist Carl Jung was the most famous modern exponent of archetypes, which he saw as "the contents of the collective unconsciousness." He felt that connecting a living person's character to a model from the past was valuable in any field:

All the most powerful ideas in history go back to archetypes. This is particularly true of religious ideas, but the central concepts of science, philosophy, and ethics are no exception to this rule. In their present form they are variants of archetypal ideas created by consciously applying and adapting these ideas to reality. For it is the function of consciousness not only to recognize and assimilate the external world through the gateway of the senses, but to translate into visible reality the world within us.[9]

Jung's archetypes were "universal images that have existed since the remotest times" found in myths and fairytales in all cultures. From a few basic psychological types, Jung believed that numerous other archetypes could arise based on family roles (father, mother, child) or approaches to life (the wise old man, the maiden, the hero, the trickster) or other personality factors. The number of possible archetypes was not fixed.[10] Archetypes in literature allow readers to identify with characters on an emotional level. The same thing happens in other arts, from painting, illustration, and sculpture to music and dance.

The movement to abolish slavery in the United States required many kinds of leaders with a variety of backgrounds and approaches to succeed. Different groups of followers—Black or white, male or female, young or old, elite or populist—identified with different leaders based on their personal qualities. Grouped together as archetypes, we can translate the main quality each type of leader brought to the abolition movement back then into what kinds of leadership might help the climate movement succeed today. So, here we'll match up famous abolitionists with famous activists from a variety of backgrounds who have been working to save the climate.

It should go without saying that no list of either abolitionists or climate leaders can be exhaustive. Even the longest list would go out of date as older people retired and new people came along. On climate alone, many notable names are missing from our list, especially professionals such as lobbyists, attorneys, or investment fund managers quietly plying their specialized skills in an insider game to sway large powerful institutions away from dirty energy and towards climate solutions. Their work is valuable and necessary to any movement to abolish fossil fuels.

But since this book is about expanding the climate movement by a factor of a hundred or a thousand to defeat the Oil Power, the list below focuses on leaders who are well-known to a broader public and who have used their fame to reach even more folks.

Green business leaders and eco-entrepreneurs from Elon Musk to Michael Bloomberg have become some of the most admired (if often controversial) heroes of climate action. That makes sense given the outsized role that trade, manufacturing, and technology play in the human activities most connected to the climate crisis, especially how we use energy. But the globalized post-industrial economy of the early twenty-first century is a very different one from the largely agrarian economy of America before the Civil War. As a result, business leaders did not play the prominent role in abolition that they must play in the battle to save the climate.

There were exceptions, especially during the Civil War when manufacturers stepped up to mass produce war materiel from uniforms to artillery shells to field rations while Wall Street bankers raised the $2 million per day that it took to keep U.S. forces in the field. Even before the war, antislavery New York landowner and philanthropist Gerrit Smith financed abolition efforts, especially as one of the "Secret Six" who bankrolled John Brown's guerilla war against proslavery forces in Kansas in the 1850s.

But overall, leaders of industry play a much bigger role in climate solutions today than they did on abolition in the nineteenth century. If anything, most business interests before the Civil War were on the side of slavery, not abolition, whether they were big Southern planters or their allies among Northern cotton manufacturers, shippers, and financiers.

It's a similar case for two other groups of experts who are and were key to each movement respectively. The climate movement started with scientists, and the abolition movement started with religious leaders. Yet, neither scientists nor religious leaders headline our list of abolition heroes.

Scientists and preachers, with their unique ability to go deep and sound the initial alarm, were crucial to spark the fire of awareness. But for both movements to catch "all on fire" as abolitionist William Lloyd Garrison put it, broad outreach to the public was required to fuel the righteous blaze. Scientists remain essential to any cause based on fighting lies with facts. Ministers of the Gospel were always important for abolition. We'll find both scientists and religious leaders on the list below, but just not at the beginning.

Finally, while trying to avoid cultural appropriation, the list employs "Hamilton" casting, matching people from then and now across race and gender to explore a deeper truth about their roles as leaders.

The Freedom Seeker: Self-Emancipated People / Residents of Frontline Communities

Enslaved people were the first abolitionists. Africans resisted enslavers' attempts to clap them into chains, herd them into inland slave pens run by slave traders and march them into slave forts built by European powers on the coast of West Africa. During the infamous Middle Passage from Africa to the Americas, when they were able, kidnapped captives staged revolts on the high seas against slave ship crews to take control of the vessel and force the crew to turn the ship around, back towards home.

Otherwise, captives went on hunger strikes or jumped overboard to liberate themselves from a life that seemed no longer worth living or to escape to an afterlife they believed would offer freedom. Once arrived on the shores of the New World, enslaved Africans and their descendants mounted the full range of resistance against their captors, from work slowdowns or stoppages, sick outs, and sabotaging equipment, to attempts to escape and, in a few famous cases, armed revolts.

Violent uprisings of the enslaved were less common in the American South than in the slave colonies of the Caribbean, though a few insurrections including those led by Nat Turner and Denmark Vesey struck fear into the hearts of white people across the South. Escape attempts became increasingly frequent once Northern states started phasing out slavery during the American Revolution.

By the middle of the nineteenth century, thousands of fugitive slaves, often aided by the secretive network of Black and white allies known as the Underground Railroad, became such big news across the country that white Americans had to take notice. After angry Southern planters succeeded in passing the draconian Fugitive Slave Act of 1850 that required every Northerner to become a slavecatcher if asked, slavery suddenly became much more real to white people across the free states. No longer safe in Northern states, fugitive slaves now had to travel all the way to Canada.

People running away from their homes and families to free themselves exposed the lie that slavery was a benign institution, as planters claimed, and forced Northern whites to consider whether a nation dedicated to human equality could much longer tolerate an economic system that treated people as property.

"Fugitive slaves pushed the nation toward confronting the truth about itself. They incited conflict in the streets, the courts, the press, the halls of Congress, and perhaps most important, in the hearts and minds of Americans who had been oblivious to their plight," writes historian Andrew Delbanco, who claims that fugitive slaves in years before Fort Sumter fired the first shots in the "war before the war."[11]

In the fight against devastation by fossil fuel companies, "frontline" or "fence-line" communities living next to coal mines, oil fields, refineries, gas compressor stations, and pipelines were the first to suffer the effects of pollution—and the first to resist. Numberless communities have fought against dirty energy projects and polluting facilities from "Cancer Alley" in Louisiana to Native American reservations in the path of the Keystone XL or Dakota Access pipelines. Nearly always low-income areas largely inhabited by people of color, frontline communities have been treated as sacrifice zones lacking the wealth and power to resist fossil fuel projects, and have suffered the most from polluted air, water, and land.

Ever since the 1970s, the champion of frontline communities has been sociologist and environmental activist Dr. Robert Bullard. Known as the "Father of Environmental Justice," for 40 years Bullard has been documenting stories of frontline communities trying to fight off dirty energy projects in speeches, articles, and 18 books on racial equity, land use planning, and environmental protection including *Dumping in Dixie: Race, Class and Environmental Quality* (2000), *Race, Place and Environmental Justice After Hurricane Katrina* (2009), and *The Wrong Complexion for Protection* (2012). In 2013, he became the first Black recipient of the Sierra Club's John Muir Award and the following year, the Sierra Club named its new Environmental Justice Award after Bullard.

Bullard has shown that the connection between slavery abolition and oil abolition is not just figurative. His research demonstrates that abolishing oil is key to completing the work begun by abolishing slavery.

His first book *Dumping in Dixie* traced a historical line from free Black communities that purchased property in Southern states after the Civil War to the polluting industries that soon followed. Ironically, sometimes oil companies bought out bankrupt cotton or sugar plantations formerly worked by the enslaved. During decades of Jim Crow segregation, descendants of slaves and their families were denied the vote, which made it easier to deprive them of education as well as water and sewer infrastructure. Since dirty industrial projects were overwhelmingly sited in their communities, descendants of emancipated slaves were also exposed to higher-than-average levels of pollutants, compromising their health and well-being for generations.

For years, Bullard's crusade to interest either mostly white environmental groups or else civil rights groups in the plight of frontline communities was a lonely one. But recently decades of persistent work ultimately paid off with much deserved attention to Bullard's concerns for frontline communities. In 2020, Bullard received a Champions of the Earth Lifetime Achievement Award for his work as an environmental justice advocate from the U.N. Environment Program.

Bullard's primary goal now is teaching young people how to advocate for change. He cautions them that "the quest for justice is no sprint. It's a marathon relay (where we must) pass the baton to the next generation of freedom fighters."[12] Youth activists occupy a couple places on this list, including the next one.

The Prophet: Frederick Douglass / Greta Thunberg

Frederick Douglass was one of the most photographed Americans of the nineteenth century, Black or white. His imposing good looks and steely gaze certainly made him photogenic. Photographers and publishers knew that there would be a market for pictures of Douglass because he had earned fame as America's leading abolitionist, advisor to Abraham Lincoln, and chief recruiter for 200,000 Black soldiers and sailors in the Civil War.

Douglas had a powerful story to tell, and he was good at telling it in different media. He gave hundreds of lectures attended by tens of thousands of mostly white Americans and published three versions of his autobiography. Along with photos sold on cards, Douglass was depicted in paintings, cartoons, and marble busts.

But Douglass preferred photos, writing that "the picture-making faculty is a mighty power" because it could make visible a person's inner nature, helping show that Black people were just as human as white ones, thus disproving claims of Black inferiority made by Southern slaveowners. Douglass also thought that images were able to build support for political movements in ways that ideas alone were not. "The few think, the many feel. The few comprehend a principle, the many require an illustration."[13]

As a person who escaped from slavery to freedom, Douglass brought personal experience of the evils of chattel slavery. But so did hundreds of runaway slaves in the North and in Canada, and dozens of them wrote and published stories of their lives under slavery. What made Douglass stand out from other self-emancipated Black Americans was rare skill as a writer and a speaker. And it was the power of his presence that made Douglass a celebrity.

Some of his most famous quotes show that Douglass was not afraid to speak in prophetic tones, calling on men and women of conscience to forsake the path of moderation and compromise to fight the grave moral ill of slavery. Consider what he said in a famous speech he gave in Rochester, New York, on July 5, 1852.

> What, to the American slave, is your Fourth of July? I answer: a day that reveals to him, more than all other days of the year, the gross injustice and cruelty to which he is a constant victim. To him, your celebration is a sham; your boasted liberty, an unholy license; your national greatness, swelling vanity...At a time like this, scorching irony, not convincing argument, is needed. O! had I the ability, and could reach the nation's ear, I would, to-day, pour forth a stream, a fiery stream of biting ridicule, blasting reproach, withering sarcasm, and stern rebuke...We need the storm, the whirlwind, the earthquake. The feeling of the nation must be quickened; the conscience of the nation must be roused; the propriety of the nation must be startled; the hypocrisy of the nation must be exposed; and the crimes against God and man must be proclaimed and denounced.

After delivering the judgment of an angry God on a sinful nation, Douglass then changed course. To inspire his audience to act, he ended the speech with a prophecy of hope: "Allow me to say, in conclusion, notwithstanding the dark picture I have this day presented of the state of the nation, I do not despair of this country. There are forces in operation, which must inevitably work the downfall of slavery...I, therefore, leave off where I began, with hope. While drawing encouragement from the Declaration of Independence, the great principles it contains, and the genius of American Institutions, my spirit is also cheered by the obvious tendencies of the age."[14]

Today, the best candidate for the role of climate prophet does not live in the United States, but she has exerted a great influence here:

Greta Thunberg. Her trademark disapproving scowl (lovable if you want to save the climate but withering if you hope to keep making money destroying it) has become as iconic as her apparent fearlessness in telling people in power right to their face how they've screwed up on stopping runaway climate disruption.

Thunberg is just as blunt as Douglass was, though she speaks not in the language of religious prophecy but in the modern world's equivalent, scientific authority: "I don't want you to listen to me, I want you to listen to the scientists," Thunberg told members of Congress while visiting the U.S. in 2019. "I want you to unite behind the science and I want you to take real action."

Thunberg doesn't need to invoke the wrath of God to deliver a hellfire-and-brimstone sermon on climate destruction. But her moral appeal is as unrelenting as any from Frederick Douglass. "The eyes of all future generations are upon you. And if you choose to fail us, I say—we will never forgive you," she warned the United Nations Climate Summit in New York on that same American trip in 2019.

She then went on to scold her audience as Douglass did about the hypocrisy of celebrating freedom in a land with millions of Americans in chains. "My message is that we'll be watching you. This is all wrong. I shouldn't be up here. I should be back in school on the other side of the ocean. Yet you all come to us young people for hope. How dare you. You have stolen my dreams and my childhood with your empty words. Yet I am one of the lucky ones. People are suffering."

A true example of the Prophet archetype, Thunberg is not afraid to stand alone in full public view to call on the sinners of the world to repent. We can remember how eighteenth-century Quaker abolitionist Benjamin Lay sat outside the meeting house on cold Sunday mornings in winter, with his shoeless foot buried in the snow to send a message with his own body—if you think I'm suffering, just think of the much worse suffering of enslaved people. Thunberg appeals to science as she whips up a jeremiad against carbon sin and climate damnation worthy of an Old Testament prophet.

Like Douglass, Greta also got her start by putting her physical presence on the line in full public view. On August 20, 2018, then in ninth grade, she started to skip school every day to stage a one-person protest outside the Swedish Parliament building in Stockholm to draw attention to the nation's failure to cut carbon emissions in line with the Paris Accord. Grasping a homemade poster reading *Skolstrejk för klimatet* (school strike for climate) she planned to hold her solitary vigil until the

Swedish elections the following month. But just before the elections, Greta announced that she would protest every Friday, coining the term Fridays for Future. Her lonely campaign soon attracted fellow protesters in Stockholm, and then spread to cities across the world, leading to two boat trips to attend gatherings of diplomats (she refuses to fly to protest the high carbon emissions of the airline industry) and a frosty brush with Donald Trump.

Right-wing media and climate deniers have not allowed Greta's youth to spare her from the meanest attacks. Patrick Moore of the Koch-funded front group the CO_2 Coalition once tweeted simply "Greta = Evil."[15]

Thunberg has made a promising start, but she's still young. Douglass was the face of Black freedom and equality spanning a career of fifty years. In today's America, the only people in climate as well known today as Douglass was in the mid-nineteenth century are Hollywood celebrities: Leonardo DiCaprio, Jane Fonda, Robert Redford, Arnold Schwarzenegger, Don Cheadle, Ed Begley Jr., and Mark Ruffalo—or Al Gore, a special kind of climate celebrity who we'll cover below.

The Publisher: William Lloyd Garrison / Bill McKibben

William Lloyd Garrison was the most famous white abolitionist. Garrison published the first issue of his antislavery newspaper the *Liberator* in Boston on January 1, 1831, with an editorial calling for the immediate emancipation of America's slaves without any compensation to slaveowners. And he wasn't going to moderate his language to avoid offending anyone.

"No! no! Tell a man whose house is on fire to give a moderate alarm; tell him to moderately rescue his wife from the hands of the ravisher; tell the mother to gradually extricate her babe from the fire into which it has fallen;—but urge me not to use moderation in a cause like the present," he wrote. Garrison rejected slavery. He also rejected compromises with owners of capital in slaves like plans for gradual or compensated emancipation. Garrison was uncompromising in putting morality ahead of money.

That's why Garrison's paper built up a national following especially among free Black people and helped revive the abolition movement after two decades of quiet. Garrison would publish continuously for the next 34 years, printing his final issue of *The Liberator* only when the Thirteenth Amendment ending slavery became law in December 1865. Garrison's paper helped develop abolitionist writers including Frederick Douglass. The archetype of an activist writer who came to lead a movement, Garrison helped found the American Anti-Slavery Society in 1833.

Publishing an abolitionist newspaper was a difficult and dangerous profession. The other major antislavery publisher of the era, Elijah Lovejoy, was killed in 1837 when a proslavery mob set fire to the warehouse in Alton, Illinois where the publisher's printing press was stored. After shooting Lovejoy five times, the mob threw the printing press out a window onto the riverbank below, hacking it to pieces and dumping them into the Mississippi River.

Garrison also faced death threats from angry slaveowners along with government sanctions from Southern states. Just seven months after the first issue of the *Liberator* was published, Nat Turner's slave insurrection broke out in Virginia. Southern leaders blamed abolitionists in general and Garrison in particular for inciting slaves to revolt. A grand jury in North Carolina indicted Garrison for distributing incendiary material and the Georgia legislature offered a $5,000 reward (equal to more than $130,000 in today's money) to arrest Garrison and bring him to the state to stand trial.

Even in Boston, the capitol of the abolitionist movement, in 1835, a proslavery mob dragged Garrison from an abolitionist meeting, tied a rope around his waist, and dragged him to Boston Common to be tarred and feathered. Garrison was saved only by Boston's mayor, a staunch abolitionist, who put the publisher in jail for his own safety. Meanwhile, the mob raised a gallows in front of Garrison's office.

"Slavery's defenders went to astonishing lengths to stifle public criticism, even in the North," explains Ken Ellingwood, author of a book on abolitionist publishers. "Slave states outlawed 'incendiary' abolitionist journals, which they said could incite slaves to rebel. Southern lawmakers sought vainly to get counterparts in the North to crack down on abolitionist materials at their source. Suppression extended even to Congress: the House of Representatives, dominated by Southerners and their Northern allies, approved a 'gag rule' in 1836 to prevent discussion of slavery within its walls."[16]

In the South, proslavery authorities even burned bags of U.S. Mail known to contain abolitionist literature, "but canny abolitionists spotted advantage in marrying the cause of the slave with freedom of the press," writes Ellingwood.

Recently, Republicans in pro-oil state governments have passed legislation that would outlaw protests against fossil fuel projects, the FBI has started to keep files on protesters, and major employers have tried to squelch criticism of their own polluting practices. As abolitionists did before them, canny leaders of the climate movement have made such attacks into an issue of free speech.

In response to a story about how Amazon threatened to fire employees who participated in demonstrations protesting the company's environmental practices, Bill McKibben tweeted, "Tell @Amazon and @JeffBezos: the world is on fire. Climate leaders don't silence employees who are sounding the alarm. This is sick behavior." And when he found out through a Freedom of Information Act inquiry that the FBI was tracking members of the climate activist group he founded, 350.org, McKibben said that the FBI's apparent failure to distinguish between nonviolent civil disobedience and domestic terrorism was contemptible.

"Trying to deal with the greatest crisis humans have stumbled into shouldn't require being subjected to government surveillance. But when much of our government acts as a subsidiary of the fossil fuel industry, it may be par for the course."[17]

So far, climate activists have not been the main targets of the kind of mobs that attacked abolitionist publishers, but leaders of the climate movement have found themselves in jail after acts of civil disobedience. The most famous of these has been Bill McKibben.

When it comes to doing for the climate what Garrison did for abolition—publishing early and often, founding a far-reaching organization, and putting his own body on the line—nobody better represents the archetype of activist publisher than McKibben. Both were and are inspiring public speakers too.

"The most important thing any individual can do to stop climate change is stop being an individual," McKibben has written. He wasn't asking people to start recycling more. He was asking them to start doing more political activism.

McKibben has published several books on the climate crisis, all with the goal of recruiting more people into the movement. In 1989, he put out the first book on the topic for a wide audience, *The End of Nature*. Thirty years later, in 2019, McKibben released *Falter: Has the Human Game Begun to Play Itself Out?* In 2008 McKibben founded the first international grassroots climate action group, 350.org. A pioneer of distributed organizing on social media, the group held more than 5,200 demonstrations in 181 countries on a single day, October 24, 2009.

More recently, McKibben has worked to draw attention to the role of banks and other Wall Street financial firms in enabling new pipelines, natural gas export facilities, and other fossil fuel projects to go forward.

"Money Is the Oxygen on Which the Fire of Global Warming Burns," was the title of an article he published in the New Yorker in 2019.[18] "What if the banking, asset-management, and insurance industries moved away from fossil fuels?" He's glad that activists are campaigning to elect greener political candidates, lobbying to pass climate legislation, taking polluters to court, and chaining themselves to the White House fence to protest new gas pipelines.

> But what if there were an additional lever to pull, one that could work both quickly and globally? One possibility relies on the idea that political leaders are not the only powerful actors on the planet—that those who hold most of the money also have enormous power, and that their power could be exercised in a matter of months or even hours, not years or decades. I suspect that the key to disrupting the flow of carbon into the atmosphere may lie in disrupting the flow of money to coal and oil and gas.[19]

That's why McKibben is drawing public attention to the investment in fossil fuels of banks, insurance companies, asset managers, and hedge funds like JPMorgan Chase, Chubb, and BlackRock. Without Wall Street money, fossil fuel companies will face their endgame, which will break their power in politics, and make serious government action on climate solutions possible.

Such work inspired the Stop the Money Pipeline campaign that we encountered above. Since several logos for subgroups of the 350.org organization that McKibben founded are listed as members of this effort, it's likely that McKibben himself played an important role. It's a pity that the website didn't make his role clearer.

Before the Civil War, Garrison also fulminated against the money power that helped make the slavedrivers of the Cotton Kingdom richer and more powerful.

Another similarity: Both Garrison and McKibben were and are early leaders of their movements who started as young men and lived to become elder statesman (or at least middle-aged statesmen). Coincidentally, the two men celebrate birthdays two days apart. Garrison was born on December 10, 1805, which made him 25 years old when he started publishing his abolitionist newspaper *The Liberator* on January 1, 1831. By the time that slavery was abolished by the Thirteenth Amendment at the end of 1865, Garrison was 60 years old.

McKibben, born on December 8, 1960, was 29 years old when he published *The End of Nature* in 1989. In 2021, when Joe Biden took office and became the first president to take serious action on climate, McKibben was the same age as Garrison was at the end of slavery, 60 years old. Let's hope the two men will also have in common that they helped lead a multigenerational movement to achieve an impossible goal, final victory.

Honorable mentions for Garrison-style advocacy applied to climate go to Naomi Klein, critic of predatory capitalism and "extreme energy" production through dangerous and dirty projects like the Alberta tar sands and offshore oil as well as pipelines like Keystone XL. Like McKibben, Klein is both an activist and an author of numerous books, including *This Changes Everything: Capitalism vs. the Climate.*

Also, investigative journalist Ross Gelbspan, historians of science Naomi Oreskes and Erik Conway, and propaganda analyst James Hoggan have helped uncover the decades-long disinformation campaign by fossil fuel companies to discredit climate science and delay action to stop climate breakdown. Another major contributor to exposing the fossil fuel campaign of lies is not a journalist but a leading climate scientist, Michael Mann, whose work we've discussed on the new ways that fossil fuel companies are trying to delay climate action.

The Statesman: Senator Charles Sumner / Vice President Al Gore

In the nineteenth century, an age before electronic media, listening to public speakers was considered entertainment comparable to watching a movie today. The abolition movement was blessed with powerful orators who held audiences in thrall with fiery speeches that sympathized with the suffering of the slave and denounced the wickedness of the slaveowner. Strong public speakers dominated the U.S. Senate too.

One of these was Charles Sumner, senator from Massachusetts. A Harvard-educated lawyer, Sumner was skilled at quoting the Bible and classical literature to sharpen piercing sarcasm aimed at the heart of his political foes. Sumner served in Congress as the leading champion for abolition for twenty years. His fame only grew during the controversy over whether new territories out West like Kansas would enter the country as free or slave states.

In 1856, Sumner gave a colorful speech on the Senate floor called "The Crime Against Kansas" comparing proslavery senators to pimps for the "harlot slavery." Proslavery Congressman Preston Brooks from South Carolina, a cousin of one of the men that Sumner had insulted, took offense. To defend the honor of his family, his state, and slaveholders across the South, Brooks decided to teach Sumner a lesson in etiquette.

As we've discussed already, the South Carolinian entered the Senate chamber, accosted Sumner at his desk, and started beating the Massachusetts senator with a wooden cane until Sumner was blinded by his own blood. Even after Sumner collapsed unconscious under the desk, Brooks kept beating him until his walking stick snapped, after which he used the cane's broken stub to beat Sumner some more. The abolitionist legislator was nearly killed in the assault, suffering head trauma, nightmares, headaches, and psychological problems afterwards.

As a sign of how polarized the country was before the Civil War over the issue of slavery, Brooks became a villain to Northerners but a hero to white Southerners, who enthusiastically sent him hundreds of replacement canes, one with the inscription "Beat him again."

Sumner was celebrated with rallies attended by thousands in cities from Boston and New York to Cleveland and Detroit. Copies of his speech were distributed across the North. Though it took Sumner three years to recover from his injuries, Massachusetts did not replace him, but instead reelected him to another term. During Sumner's convalescence, the state was proud to be represented by his empty chair in the Senate chamber, to protest slavery and make a statement in favor of free speech. As it was with abolitionist newspaper publishers, so it was with senators who agitated to deprive slaveowners of their money and power: the threat of physical violence was an occupational hazard.

When Sumner returned to the Senate in 1859, his fellow Republicans advised him to tone down his criticism of slaveowners. Sumner refused, replying that "When crime and criminals are thrust before us, they are to be met by all the energies that God has given us by argument, scorn, sarcasm, and denunciation." Accordingly, the first speech he delivered in the Senate after his return, given two months after Fort Sumter, "The Barbarism of Slavery," positioned the Civil War not as a conflict between diverging sections of the country over the fate of the Union but as a battle between civilization and barbarism over the fate of slavery.

> On the one side are women and children on the auction-block; families rudely separated; human flesh lacerated and seamed by the bloody scourge; labor extorted without wages; and all this frightful, many-sided wrong is the declared foundation of a mock commonwealth.
>
> On the other side is the Union of our Fathers, with the image of "Liberty" on its coin and the sentiment of Liberty in its Constitution, now arrayed under a patriotic Government, which insists that no such mock Commonwealth, having such a declared foundation, shall be permitted on our territory, purchased with money and blood, to impair the unity of our jurisdiction and to insult the moral sense of mankind.[20]

Once the Civil War started, Sumner continued to give incendiary speeches to push the federal government to transform the U.S. war effort from a fight to save the Union into a fight to free the slave. Though Lincoln originally hesitated to emancipate all slaves to avoid the risk of pushing slave states still in the Union—the border states of Delaware, Maryland, Kentucky, and Missouri—into the Confederacy, Sumner and the other Radical Republicans in Congress kept pressuring Lincoln to free slaves wherever and whenever fighting and military occupation made possible.

As a compromise between Lincoln and Radical Republicans in Congress led by Sumner, Congress passed and Lincoln signed two Confiscation Acts in 1861 and 1862, orders to free slaves directly employed in the Confederate war effort.

Here, emancipation was promoted to the Northern public—an electorate with many white voters including so-called Free Soilers and War Democrats who opposed full abolition—as a military necessity to help win the war. In 1863, the Emancipation Proclamation followed, again, undertaken not explicitly as a step to abolish slavery but instead merely to deprive the enemy's economy of the resource of slave labor and make Black troops available to the United States. But the effect was the same: enslaved Southerners got their freedom. And such incremental moves presented as war measures helped prepare the Northern public gradually for the final push for abolition.

That push came the following year when Sumner submitted a proposed Constitutional amendment to abolish slavery throughout the nation: "All persons are equal before the law, so that no person can hold another as a slave." The Senate wound up passing less expansive language as the final Thirteenth Amendment: "Neither slavery nor involuntary servitude, except as a punishment for crime whereof the party shall have been duly convicted, shall exist within the United States, or any place subject to their jurisdiction."

Charles Sumner suffered violence and nearly died in service of abolition. Until January 2021, it appeared that Capitol Hill in the twenty-first century was going to be less violent than it was in the nineteenth century. And fortunately, the senior statesman of the climate movement, former Vice President Al Gore, wasn't physically assaulted by an oil-state legislator when he served previously in the Senate.

Yet, during a career as an oil abolitionist as long as Sumner's was as a slavery abolitionist, Gore has faced his share of harsh political attacks. "Gore has for years been pilloried by the right-wing press, including Fox News and the rest of the Murdoch media empire, for the size of his home, his electricity bills, and even his weight."[21]

Gore is as famous for serving as Bill Clinton's vice president as for his post-government service giving the world's most famous PowerPoint presentation, *An Inconvenient Truth*. After it came out as a documentary film in 2006, attacks on Gore from fossil fuel PR operatives started flying fast and furious.

Dubbed "America's First Political Climate Action Hero," Gore got started in politics young, elected to represent Tennessee in the House of Representatives at the age of 28.[22]

Just as Sumner was early to the fight against slavery in the Senate, Gore was also early to the fight for a livable climate on the national stage. Inspired by work measuring carbon dioxide in the atmosphere by a professor during his college days, Harvard oceanographer and climate researcher Roger Reveille, Gore held the first ever hearing in Congress on climate change in 1976. Even while advancing in his political career, winning election to the Senate in 1984 and running for president in 1988, Gore found time to work on his first book *Earth in the Balance: Ecology and the Human Spirit*, which became a bestseller.[23]

Elected vice president to Bill Clinton in 1992, Gore pushed hard for a carbon tax but was thwarted by Republicans in Congress beholden to the Oil Power. During the Civil War, Charles Sumner helped the Lincoln administration to keep England and France from joining the conflict as an ally of the Confederacy. Gore performed diplomacy on behalf of climate, negotiating the first international global warming treaty in 1997, the Kyoto Protocol. Unfortunately, this became just another instance of fossil fuel companies working successfully to delay climate action when they got Republicans to block the treaty's ratification by the Senate.

In the famous disputed presidential election of 2000, the Supreme Court decided in favor of George W. Bush over Gore, depriving the United States of what would have been our first climate president. Who knows how history would be different today if Gore had been able to push for serious climate action like a carbon tax, an end to oil subsidies, or a wartime-style mobilization for clean energy from the White House? It is certain that with Gore's loss the country and the world that counts on American leadership lost two decades in the fight against climate catastrophe.

Just as Sumner did not soften his zeal for abolition when he returned to the Senate in 1860 after his brush with death at the hands of a proslavery assailant, so Gore did not drop out of the fight to save the climate after his political defeat by Texas oilman George W. Bush.

If anything, Gore became more famous as a climate activist outside of government, turning the slideshow he had given more than a thousand times over the years into a documentary film, *An Inconvenient Truth*. No ordinary environmental film, it became the third highest earning documentary ever and won two Academy Awards. Gore followed up on the film's massive success by publishing *The Assault on Reason*, ruthlessly excoriating politicians Charles Sumner-style for ignoring the clear facts of climate science and dithering about taking action to save civilization. In 2017 Gore released a revised version of the book to cover social media and the election of Donald Trump.[24]

Gore's film and book proved so popular and influential both inside the U.S. and around the world that he shared a Nobel Peace Prize in 2007 with the Intergovernmental Panel on Climate Change for "efforts to build up and disseminate greater knowledge about man made climate change, and to lay the foundations for the measures that are needed to counteract such change."[25]

In his acceptance speech, Gore called for society to stop the pollution from fossil fuels that was the equivalent of waging war on the earth. Then he urged countries to act on climate change as if they were mobilizing for a war. "For now, we still have the power to choose our fate," Gore appealed to his audience. "The only question is this: have we the will to act, vigorously and in time, or will we remain imprisoned by a dangerous illusion?"[26]

Gore continued to come up with new ways to draw attention to climate breakdown and build support for solutions, branching out into two new areas. Like McKibben, Gore started his own international organization, the Climate Reality Project. Founded in 2006, the same year that *An Inconvenient Truth* was released in theaters, Gore's group has trained more than 31,000 speakers to mobilize citizens in 170 countries to become activists for clean energy.

A couple years earlier in 2004, with Goldman Sachs veteran David Blood, Gore co-founded Generation Investment Management. With more than $31 billion in assets as of 2021, the firm invests only in companies that are both financially and environmentally sustainable. By delivering the "superior investment performance" that the firm promises its clients, Gore hopes to prove that businesses "can make more money if they change their practices in a way that will, at the same time, also reduce the environmental and social damage modern capitalism can do."[27]

Like a modern-day Charles Sumner, Gore has continued to give moving speeches to warn Americans about the threat of global heating and to recruit citizens to fight for climate solutions. In 2016 Gore gave a TED Talk, "The Case for Optimism on Climate Change," that has gotten more than 2.2 million views. Despite the title, Gore opened with some decidedly unoptimistic warnings to answer the question "Do we really have to change?" As temperatures have risen along with climate pollution, weather disasters have become more apocalyptic: biblical rainstorms and flooding in New York City, Tucson, Houston, Chile, and Spain along with heatwaves and drought that brought wildfires to the western U.S. and historic crop failures and civil unrest to Syria. Not to mention big pieces of ancient ice breaking off glaciers in Greenland and Antarctica. "Every night on the TV news now is like a nature hike through the Book of Revelation," Gore said in his talk.

In the manner of the most popular abolitionist orators, after pulling his audience down into a valley of despair, Gore then lifted them back up onto hills of hope. Clean energy is growing fast, he pointed out: More solar panels and wind turbines have been installed than anybody predicted, and they're being backed up by energy storage from batteries that have continued to get cheaper every year. Meanwhile, dirty energy is dying. New coal plants were cancelled, and old ones are closing. China has committed to ambitious carbon-cutting goals and Americans are marching in the streets to demand climate action.

Gore ended on a note of moral fervor not merely reminiscent of abolition, but with an explicit reference to the historic movement to emancipate slaves:

> We now have a moral challenge that is in the tradition of others that we have faced. One of the greatest poets of the last century in the U.S., Wallace Stevens, wrote a line that has stayed with me: "After the final 'no,' there comes a 'yes,' and on that 'yes,' the future world depends." When the abolitionists started their movement, they met with no after no after no. And then came a yes...When any great moral challenge is ultimately resolved into a binary choice between what is right and what is wrong, the outcome is foreordained because of who we are as human beings. Ninety-nine percent of us, that is where we are now, and it is why we're going to win this. We have everything we need. Some still doubt that we have the will to act, but I say the will to act is itself a renewable resource.[28]

Honorable mention to fill the archetypal role of Statesman goes to other climate leaders with long, distinguished careers at the top levels of state and federal government. Biden administration Special Presidential Envoy for Climate John Kerry negotiated the Paris Agreement in 2016 and signed the U.S. back into the global climate compact in 2021.

Rhode Island Senator Sheldon Whitehouse delivered his "Time to Wake Up" speech in Congress 279 times over 12 years. Governor Jay Inslee came up with one of the most famous quotes of the climate movement: "We are the first generation to feel the sting of climate change, and we are the last generation that can do something about it." His work in Washington State inspired much of Biden's own climate plan including its ambitious goal of zero-emission electricity generation across the U.S. by 2035. Ever since he served as governor of California from 2003 to 2011 Arnold Schwarzenegger has been perhaps the most outspoken Republican in favor of action against climate disruption. He has called climate the issue of our time and has tirelessly urged politicians to stop treating climate as a partisan political issue but to unite across party lines to implement solutions.

The Radical: Rep. Thaddeus Stevens / Rep. Alexandria Ocasio-Cortez

"He grew up poor, made a fortune and from then on championed the weak and any other group who wasn't able to fight equally," wrote singer Bob Dylan about one of the figures from history he most admired, Civil War-era Congressman Thaddeus Stevens. "He got right in there, called his enemies a 'feeble band of lowly reptiles who shun the light and who lurk in their own dens...' [He] could have stepped out of a folk ballad."[29]

No member of Congress better embodied the archetype of the Radical than Pennsylvania's Stevens, a more junior member of government than the Statesman with a more populist style and a more aggressive program.

Back then, we should recall, the positions of the two American political parties on racial issues were reversed. The Democrats were the proslavery party, with big support among Southern planters, and the Republicans were the antislavery party, or at least the party that opposed the extension of slavery into new areas. As with political parties today, most members were moderate and interested in compromise with the other side to get laws passed faster and more easily, even if that meant watering down those laws.

Other members of the same party were less willing to compromise and insisted on holding fast to their ideals to fight for more significant change. Thaddeus Stevens was in the second group. His deep-seated hatred of slavery and visionary support for equal rights for Black Americans put him on the far left of the left-leaning party of his day, a group known at the time as the Radical Republicans.

Born in poverty with clubfoot that left him with a permanent limp, Stevens taught school and practiced law before getting involved in politics at the local and state levels as an advocate for free public education and abolition. As an attorney he defended Black clients accused of being fugitive slaves. In the Pennsylvania state legislature, Stevens fought against disenfranchising Black voters in localities where they were allowed to vote. Like many abolitionist leaders, Stevens was a powerful public speaker with a strong moral appeal. His use of oratory inspired by sermons was captured perfectly by actor Tommy Lee Jones who portrayed Stevens as a moralizing curmudgeon who could play political hardball, when necessary, in Steven Spielberg's 2012 film *Lincoln*.

In real life, Stevens wasn't afraid to ham it up in the cause of abolition. "I wished that I were the owner of every Southern slave, that I might cast off the shackles from their limbs and witness the rapture which would excite them in the first dance of their freedom," he told Pennsylvania lawmakers in 1837. A resident of Lancaster, Pennsylvania, located just north of the Mason-Dixon line, Stevens risked his standing in society and his ability to hold public office by breaking the law: He served as a volunteer stationmaster on the Underground Railroad, coordinating the movement of fugitive slaves and offering them refuge in a hidden cistern connected to his house by a secret passage.

When Stevens was elected to Congress in 1858, he joined Senator Charles Sumner in opposing the expansion of slavery and resisting concessions by the federal government to the demands of slaveowners. Three years later in 1861 when fighting broke out in the Civil War, like Sumner, Stevens argued that the federal government should not allow slavery to survive the conflict.

He was frustrated by the slowness of Abraham Lincoln to use his war powers to emancipate slaves in the South. "Our object should be not only to end this terrible war now, but to prevent its recurrence. All must admit that slavery is the cause of it. Without slavery we should this day be a united and happy people...The principles of our Republic are wholly incompatible with slavery."[30]

But abolition by itself wasn't good enough for Stevens, who also wanted the federal government to confiscate land from rebellious planters and distribute it as small farms to newly freed slaves, a plan that became known as "40 acres and a mule."[31] Then he wanted Black men given the right to vote and federal troops to remain stationed in the South after the war to enforce that right by force.

Stevens's views were well known by white Southerners, who hated the Pennsylvania abolitionist for his fiery attacks on the source of cotton planters' wealth. Southern leaders hated Stevens so much that

Confederate troops crossing into Pennsylvania in the campaign culminating in the Battle of Gettysburg attempted to capture the abolitionist congressman not once but twice. When asked if would take Stevens to Libby Prison in Richmond, which housed Union prisoners of war in crowded and squalid conditions, Confederate General Jubal Early, a brutal supporter of slavery and white supremacy who led the raids to capture Stevens, replied that he would have hanged Stevens and divided his bones among the Confederate states.[32]

Famous for his acerbic wit, Stevens was even more insulting towards slaveowners and their Northern enablers than Charles Sumner. Hated by his political opponents, Stevens was eminently quotable. Campaigning for the Republicans in Lancaster in 1856, Stevens attacked the Democratic presidential nominee, a proslavery Northerner from Stevens' own hometown of Lancaster: "There is a wrong impression about one of the candidates. There is no such person running as James Buchanan. He is dead of lockjaw. Nothing remains but a platform and a bloated mass of political putridity."[33]

A few years later, after the armed attack on Harpers Ferry that made him famous in the North and infamous in the South, Stevens taunted Southern leaders who took pride in their military ability when he quipped that "John Brown deserves to be hung for being a hopeless fool! He attempted to capture Virginia with seventeen men when he ought to know that it would require at least twenty-five."

In his first speech in Congress, Stevens showed that he came not to persuade, but to taunt the Slave Power. On the House floor, Stevens dared Southern whites who claimed that slavery was benign to prove it by making slavery voluntary. Then, if slavery was so attractive, planters should just sit back and watch their enslaved workforce grow. Southern leaders, who banned "incendiary" abolitionist literature from the U.S. Mail in Southern states for supposedly inciting slave insurrections, must have been especially insulted by Stevens's sarcasm:

> Gentlemen on this floor and in the Senate, had repeatedly, during this discussion, asserted that slavery was a moral, political, and personal blessing; that the slave was free from care, contented, happy, fat, and sleek. Comparisons have been instituted between slaves and laboring freemen, much to the advantage of the condition of slavery. Instances are cited where the slave, having tried freedom, had voluntarily returned to resume his yoke.

Well, if this be so, let us give all a chance to enjoy this blessing. Let the slaves, who choose, go free; and the free, who choose, become slaves. If these gentlemen believe there is a word of truth in what they preach, the slaveholder need be under no apprehension that he will ever lack bondsmen. Their slaves would remain, and many freemen would seek admission into this happy condition...Nor will we rob the mails to search for incendiary publications in favor of slavery, even if they contain seductive pictures, and cuts of those implements of happiness, hand-cuffs, iron yokes, and cat-o'-nine-tails.[34]

Stevens was ahead of his time. His plan for full civil rights for freed people guaranteed by strong federal enforcement was considered too extreme by most white Americans in his day, including most of his fellow Republicans. The kind of racial equality that Stevens preached would have to wait nearly a century until the civil rights movement of the 1950s and 1960s was able to win the same demands.

One of the fieriest abolitionists to ever achieve public office, Stevens spared no zeal in pushing not only for emancipation but for full racial equality, no matter how unpopular his ideas were to most Northerners at the time. "There can be no fanatics in the cause of genuine liberty," as he put it.[35]

Nobody embodies the Radical archetype for climate solutions like Alexandria Ocasio-Cortez, firebrand congresswoman from New York, champion of the Green New Deal, and recruiter of young people to climate activism. Like Thaddeus Stevens, she came from a marginalized background. In her case, both her parents were immigrants from Puerto Rico. Her father became an architect, which made her family more prosperous, but his death in 2008 brought financial worries. After graduating cum laude from Boston University, she moved back home to the Bronx and took a job as a bartender to help her mom who cleaned houses and drove school buses.

Ocasio-Cortez got into politics with Bernie Sanders's presidential primary campaign in 2016. After the presidential election, she took a cross-country road trip with stops to give speeches and meet with activists about the water crisis in Flint, Michigan, and the campaign to stop the Dakota Access Pipeline on the Standing Rock Indian Reservation in North Dakota. Impressed by the determination of anti-pipeline activists assembled from across the country, Ocasio-Cortez changed her views about what it took to succeed in politics.

"I felt like the only way to effectively run for office is if you had access to a lot of wealth, high social influence, a lot of dynastic power, and I knew that I didn't have any of those things." But the activism at Standing Rock convinced her that ordinary people could become leaders too.[36]

The day after her visit to Standing Rock, based on a nomination submitted by her brother after the Sanders campaign, she got a call from a group called Brand New Congress which was recruiting more diverse and progressive candidates to run for office.

It was as a bartender that Ocasio-Cortez first ran for Congress, also sharing with Thaddeus Stevens a background outside of professional politics. With grassroots mobilization and support from the left, in the Democratic primary election she defeated a powerful ten-term incumbent, Joe Crowley, and then went on to handily win the general election to enter Congress at age 29 as the youngest member of the House.

Even before she was sworn in, Ocasio-Cortez took up the mantle of Congressional Radical. Just like Thaddeus Stevens, who criticized Abraham Lincoln for moving too slowly on emancipation, Ocasio-Cortez clashed with leaders of her own party.

On the first day of orientation for new representatives, she staged a sit-in for climate action outside the office of House Democratic Leader Nancy Pelosi. Backed up by 200 young activists from the Sunrise Movement, Ocasio-Cortez said that the action was not about conflict but friendly persuasion, meant to encourage Pelosi to put climate action front-and-center in the coming session of Congress.

Word had gotten around that Democratic leaders in Congress had compiled a list of a half dozen or more issues they planned to focus on when they took over the House of Representatives as the new majority after triumphing in the 2018 midterm elections, but that climate change was not on the list. AOC and Sunrise took preemptive action to convince Pelosi that climate should not be ignored as Democratic leadership had done in the past.

"This is not about me, this is not about the dynamics of any personalities," she told reporters outside Pelosi's office. "But this is about uplifting the voice and the message of the fact that we need a Green New Deal, and we need to get to 100 percent renewables because our lives depend on it."[37]

Once she took office, Ocasio-Cortez, like Bernie Sanders a member of the Democratic Socialists of America, pushed quickly for radical action on both climate solutions and populist economic stimulus. Describing climate change as "the single biggest national security threat for the United States and the single biggest threat to worldwide industrialized civilization," in February 2018, she teamed up with Senator Ed Markey to introduce a resolution in both houses of Congress supporting a Green New Deal.

AOC gave new life to a concept that had been kicking around progressive circles for more than a decade to entwine aggressive action to cut climate pollution with serious government spending to boost the economy and promote fairness.

Calling for trillions of dollars of investment by the federal government over a ten-year period, the GND would decarbonize the economy; create and guarantee jobs across a new clean economy; and ensure a "just transition" that would avoid the racial inequities of the original New Deal which left Black Americans out of many programs. The GND should offer workforce development and job guarantees, along with strong labor, environmental, and nondiscrimination standards for "low-income communities, communities of color, indigenous communities, [and] the front-line communities most affected by climate change, pollution, and other environmental harm," in the words of Ocasio-Cortez's resolution.[38]

In the 1860s, Radical Republican Thaddeus Stevens pushed to transform American aims in the Civil War from saving the Union to ending slavery, both for moral reasons and for a very practical reason, to prevent future sectional conflict between the North and the South. Today, Democratic Socialist Alexandria Ocasio-Cortez is also pushing to make the fight to save the climate about ending poverty and pollution for both moral and practical reasons. Not only is it a moral question to ensure fair access to all Americans regardless of race or wealth to a clean environment and the chance to prosper. It's practical too.

Making it impossible for any group of people to suffer more than anyone else is also the best way to make sure the problem of climate is permanently solved. If there are no sacrifice zones, no communities where it's OK to dump coal ash or build gas pipelines, then in the future it will be much harder, if not impossible, to keep digging up, pumping, and burning fossil fuels.

Just like Thaddeus Stevens, AOC's willingness to confront the opposition has made her bitter enemies among reactionaries while creating tension with leading Democrats. Though a junior member of Congress, Ocasio-Cortez has gotten more attention in the media than most presidential candidates, becoming one of the most recognizable faces of her party and the climate movement.

As Thaddeus Stevens was public enemy number one for slaveholders before and during the Civil War, so recently AOC has become the leading target of a right-wing largely financed by Big Oil and stoked by Fox News. A study of a single six-week period in 2019 found that Fox and its sister station Fox Business mentioned Ocasio-Cortez 3,181 times, an average of 76 times a day.

Attacking her advocacy for a Green New Deal, Fox personalities have denounced AOC as a hypocrite for using cars while belittling her for having worked as a bartender and dismissing her for being young. Sean Hannity called her "the real speaker of the House,'" instead of Nancy Pelosi. Tucker Carlson has called her an "idiot wind bag," a "pompous little twit," a "fake revolutionary," "self-involved and dumb," a "moron and nasty and more self-righteous than any televangelist."[39]

Apparently, Carlson was immune to the irony of how these terms might apply more to himself than anyone else.

Such invective certainly has stoked anger on the right and may have encouraged pro-Trump insurrectionists who attacked the U.S. Capitol on January 6, 2021, to specifically seek out AOC, hoping to kidnap and possibly kill her. Attackers loudly banged on the doors leading to her office shouting "Where is she? Where is She?" Ocasio-Cortez recalled her reaction: "This was the moment where I thought everything was over. I thought I was going to die."[40]

Fortunately, AOC escaped the MAGA mob unharmed, no thanks to Fox News and those on the right who had stoked right-wing ire against her—including prominent climate deniers, as we saw—largely for her work to raise the alarm on climate breakdown and promote the Green New Deal. The archetype of the Radical is no stranger to death threats. What happened to AOC in 2021 is eerily reminiscent of the violence by Confederate raiders in 1863 directed at Thaddeus Stevens because of his strong stance for abolition and civil rights. Serving as the Radical is a test of courage whether your mission was to abolish slavery in the nineteenth century or is to abolish oil today.

AOC is still at the beginning of what anyone who cares about the climate must hope will be a long and productive political career. Thaddeus Stevens, who was born in 1792 while George Washington was president, got a late start in national politics, first winning election to Congress at age 57. Yet, Stevens still managed to remain in

Congress, on and off, for the better part of two decades, pushing hard for abolition and equal rights the whole time. Perhaps because she's young and obviously has bright prospects in politics is why Fox News has attacked AOC so vociferously, to preemptively destroy her political advancement, just as the conservative TV network focused on Hillary Clinton for decades before she ran for president.

But attacks from the right will only make AOC more popular with most of the public that supports climate action, just as attacks by Confederates bolstered support in the North during the Civil War and Reconstruction for Thaddeus Stevens. Ocasio-Cortez got a head start of nearly 30 years on Stevens and she could serve in national office for decades to come. So far, she has managed to pack a massive amount of hard fighting for climate solutions and social equity into a brief career in Congress. If AOC can keep up the momentum, she's sure to play as a heroic a role in abolishing oil as Stevens did in abolishing slavery.

The Bridge Builder: Rep. Cassius Clay / Former Rep. Bob Inglis

Like Stevens, Cassius Clay also served in the House of Representatives before the Civil War, but his background gave him a different role. Clay was the rare abolitionist from a slaveholding family in a slave state. Born to planters in Kentucky, Clay learned to hate slavery while away at college at Yale, after hearing a lecture by William Lloyd Garrison (you can understand why Southern planters didn't want to send their sons to college in New England).

Clay went on to found the Wide Awakes, a paramilitary group to support Lincoln and other Republican candidates in the elections of 1860. This is the same Wide Awakes that served as the inspiration for today's Sunrise Movement Wide Awakes, young activists for climate solutions, and we'll learn more about the original Wide Awakes below.

Meanwhile, after Confederates fired on Fort Sumter and started the Civil War, Clay organized a troop of militia to defend Washington, DC from attack both by rebel forces in Virginia and disloyal troops in the city's own militia unit, the National Rifles, known to be infested with Confederate sympathizers.[41]

In January 2021, another former Republican congressman from a Southern state stood up against insurrectionists trying to topple the federal government. Bob Inglis served as U.S. Representative for South

Carolina's Fourth Congressional District from 1993 to 1999 and again from 2005 to 2011. Just after the Capitol insurrection, Inglis published an op-ed denouncing the mob and calling out leaders of his own party for inciting the violence. "Oh, that my party had taken a path illuminated by truth rather than one darkened by useless conspiracy theories. Oh, that my party had chosen more leaders of character willing to speak truth to our own people. Oh, that we might learn from our mistakes and bear the fruit of repentance."

In the piece, Inglis explained that during his time in Congress while attending political rallies, he had several opportunities to egg on angry Tea Party crowds against President Barack Obama. During one demonstration to protest Obamacare in 2009 held in Washington, Inglis was sure that the crowd could have been incited to storm the Capitol building with the wrong leader and the right spark. But his conscience wouldn't permit Inglis to stoke what he felt like was unjustified and dangerous partisan rage.

A Republican operative told Inglis to play nice with the Tea Party if he wanted to remain in office, but Inglis rejected this advice. As a result, he was defeated in a 2010 Republican primary and replaced by a Tea Party opponent who attacked Inglis for, among other things, supporting legislation to cut climate pollution.

"The Tea Party didn't want to hear civility, and they sure as hell didn't want to hear my message about climate change in the 2010 cycle. (Thankfully, that latter part has changed. As the sea level rises and septic tanks stop working, even the most ardent disputer of science realizes that he or she has a problem!) I lost an election in 2010, but I didn't lose my soul. And I didn't fan the flames that blew up into an insurrection at the U.S. Capitol on Wednesday."[42]

Since he left Congress, Inglis has worked to build bridges across party lines not only for the peaceful operation of government but also for climate solutions. A convert from climate denial, Inglis, a self-described pro-life conservative, has become one of America's best-known apostles to conservative voters on the threat of climate disruption. In opposition to Republican leaders who've marched in lock step against even the mildest climate solutions on the federal level, Inglis has continued to believe that rooftop solar power will appeal to conservatives who value the free market, energy independence, and self-reliance.

To recruit conservatives as climate advocates, Inglis founded a group called RepublicEN, "the conservative answer to climate change," to promote clean energy and free-market climate solutions. Inglis was the recipient of the 2015 Profile in Courage Award from the John F. Kennedy Library Foundation "for the courage he demonstrated when reversing his position on climate change after extensive briefings with scientists, and discussions with his children, about the impact of atmospheric warming on our future." Inglis "featured prominently" in the 2014 documentary *Merchants of Doubt*, exposing the tactics of climate-science deniers.[43]

The General: Harriet Tubman of the Underground Railroad / Varshini Prakash of Sunrise

Harriet Tubman acquired the nickname "The General" in her early life for her work leading enslaved people to freedom on the Underground Railroad, even before she became the first woman to command U.S. Army troops in battle during the Civil War. Before the war, she rescued 300 people from slavery and "never lost a passenger," as she put it. Whether an official army commander or just a leader of an organized and well-trained force, the General is ready for confrontation on the battlefield of war or politics.

"I have heard their groans and sighs, and seen their tears, and I would give every drop of blood in my veins to free them," Tubman said.[44] She combined a compassionate heart with nerves of steel. After escaping from slavery on the Eastern Shore of Maryland in the fall of 1849, Tubman connected with abolitionist leaders in Philadelphia and then with the secret network of Black and white people who operated the Underground Railroad.

As a petite Black woman, Tubman was the least protected type of person in American society, and the most vulnerable to violence. This did not stop her from making thirteen trips back to Maryland to rescue enslaved members of her family and others held in bondage. On these trips, she showed both the courage of a General and the strategic forethought of a battlefield planner, using stealth to maximize her advantages and minimize the advantages of the enemy.

Tubman insisted on traveling only at night and carried a loaded pistol for protection, navigating by the stars while trying to avoid the dangers of the woods in the dark. Threats ranged from slave catchers and their bloodhounds to copperhead snakes, coyotes, and wolves; from exhausting treks through swamps and over mountains to panicked fugitives who would threaten to scream or bolt, exposing the whole group to capture. When she travelled by train to get closer to the plantations where she could meet her next group of enslaved people ready to escape, she would disguise herself as an old woman or even as a man to throw bounty hunters off track.[45]

Tubman wound up living to age 91, but in her years on the Underground Railroad and as a commander of Black Union raiders during the Civil War, she was willing to give her life for a cause she believed was just. Echoing Patrick Henry's famous line "Give me liberty or give me death," Tubman put herself squarely in the line of American patriots. "There are two things I've got a right to, and these are, Death or Liberty—one or the other I mean to have. No one will take me back alive; I shall fight for my liberty, and when the time has come for me to go, the Lord will let them kill me."[46]

Today, the General archetype is filled by another young woman of color, the Sunrise Movement's Varshini Prakash, who came into public prominence when she led the sit-in at House Speaker Nancy Pelosi's office in 2019 with AOC. In 2017, Prakash co-founded a group to enlist young activists in electing candidates who push for the Green New Deal and other climate solutions and even to oust candidates who continued to take money from Big Oil.

"The Sunrise Movement is a youth movement to stop climate change and create millions of good jobs in the process," says the group's website. "We're building an army of young people to make climate change an urgent priority across America, end the corrupting influence of fossil fuel executives on our politics, and elect leaders who stand up for the health and wellbeing of all people."

Like AOC, Prakash was a supporter of Bernie Sanders for president. Prakash was a strong critic of Joe Biden, and during the 2020 primary campaign, the Sunrise Movement graded Biden's climate plan as an F minus. Yet, when Biden agreed with Sanders to form a joint task force on climate change in 2020, Prakash accepted Sanders's invitation to serve alongside Biden appointees.

A true General, Prakash has shown courage to stand up to powerful officials and strategic thinking enough to guide her on effective campaigning and building unlikely alliances to achieve success.

The Storyteller: Harriet Beecher Stowe / Cli-Fi Authors

Originally titled *Uncle Tom's Cabin; or the Man that Was a Thing*, Harriet Beecher Stowe's novel about slaves and their masters became both the bestselling book of the nineteenth century and the top-selling work in American literature all the way up to the present day. In its first year, Stowe's novel sold more than 300,000 copies in the U.S. and more than 2 million in other countries, with translations in 37 languages, including three editions in Welsh, outselling any book on earth at the time except the Bible.

Tolstoy praised *Uncle Tom's Cabin* as the highest example of art. Royalties of $10,300 for three months of sales in 1852 were, according to the New York Times, "the largest sum of money ever received by any author, either American or European, from the actual sales of a single work in so short a period of time."

In a time before fan fiction, *Uncle Tom's Cabin* spawned numerous spinoffs in literature, theater, and song. This included at least 14 proslavery novels including John W. Page's *Uncle Robin, In His Cabin in Virginia, and Tom without One in Boston*, published the year after Stowe's book came out.

But this powerful activist novel, especially appealing to female readers before the Civil War, went out of fashion in the twentieth century. Art novelists and academic literary critics (who were mostly male) rejected Stowe's "sentimental" style as low-quality popular writing unfit to share the canon of great works defined by male authors like Hawthorne, Melville, Emerson, Thoreau, and Whitman. And civil rights activists denounced Stowe for giving the world the character of the long-suffering Uncle Tom, whose name has become an insult for Black people who collaborate with white racists.

However modern literary tastes and attitudes about race respond to *Uncle Tom's Cabin*, its importance for us is that Stowe's book had massive influence in America and across the world in its time and helped build support for antislavery by preaching far beyond the abolition choir. In fact, people never stopped reading Stowe's novel. It's remained a best seller in the U.S. and abroad, with more than 600 editions published in English and in translation (including 17 in Hebrew). Contemporary literary critics, including Henry Louis Gates who put out an annotated edition of *Uncle Tom's Cabin* in 2007, now say that Stowe's novel was both good activism and good literature.[47]

Stowe's novel came out it in 1851 and capitalized on Northern anger over the infamous Fugitive Slave Act passed the previous year that required every Northerner, if called upon by the authorities, to assist in the capture of fugitive slaves. The book galvanized support for abolition. A possibly mythical story attests to the belief among Americans of the last 150 years that *Uncle Tom's Cabin* was key to enlisting public opinion on behalf of enslaved people. When Stowe visited the White House in 1863 as the Civil War raged, Lincoln reportedly said "So you're the little woman who wrote the book that started this great war."

Like Tubman, Stowe was friends with Frederick Douglass. She based some of her descriptions of slavery on his writings, and Douglass, for his part, praised Stowe, comparing her to Shakespeare and Robert Burns and praising *Uncle Tom's Cabin* as "the *master book* of the nineteenth century."[48]

Stowe's innovation in activism was to appeal to the heart, rather than the head, and thus reach a much larger public than abolitionists could reach through pamphlets, lectures, or even illustrations and photos. As a storyteller, Stowe placed memorable characters, from the saintly Tom to the little girls Topsy and Eva to the dastardly slaveowner Simon Legree, into a plot with all the elements of a page-turner. Readers enjoyed the drama, suspense, horror, and tragedy of Stowe's story balanced by humor, tenderness, and ultimately, the triumph of good over evil. Stowe appealed well to the most avid novel readers of her society, white women. That turned out to be a smart move strategically to build wider support for abolition.

Women could see the horrors of treating humans as property when it disrupted the female sphere of the family and domestic life. The story of *Uncle Tom's Cabin* was less about the suffering of individuals than it was about the tragedy of loving homes torn apart by the buying and selling of Black people by white owners who needed money or labor. No matter how kind a slave master might be, in the end, the accidental cruelty of the domestic slave trade wound up separating mothers from children, wives from husbands, and brothers from sisters.

By helping her white readers identify with the enslaved families in the book, Stowe generated sympathy across racial boundaries, a feeling which in turn generated support for abolition. As one historian put it, "No abolitionist argument proved more compelling than that testifying to the conflict between slavery and domesticity" found in *Uncle Tom's Cabin.*[49]

Today, in a more fragmented publishing and cultural market, the Storyteller archetype for climate remains to be fully realized. There are many able writers of climate fiction, known as "cli fi," from literary fiction writers including Octavia Butler, Margaret Atwood, Barbara Kingsolver, and TC Boyle to science fiction writers like Kim Stanley Robinson and Paolo Bacigalupi. But none of their books has risen to the level of popularity or influence enjoyed by *Uncle Tom's Cabin*. Perhaps it will take a young adult book, like a climate-change version of *Harry Potter*. Or in a popular culture less dependent on books than movies, it might take a climate-focused film—that's about people and not science—with the reach of *Star Wars*.

The Poet: Julia Ward Howe / Amanda Gorman

Early in the Civil War, poet Julia Ward Howe gave birth to what became the anthem of U.S. troops in the Civil War, a song that civil rights leader and historian W.E.B. DuBois called "the noblest war song of the ages."[50] "The Battle Hymn of the Republic" put new words to the tune of a popular marching song, "John Brown's Body."

One version of that song was explicitly abolitionist, starting out: "John Brown's body lies a moldering in the grave,/While weep the sons of bondage, whom he ventured all to save;/But, tho' he lost his life in struggling for the slave,/His Soul is marching on."[51] This version appealed to hardcore antislavery activists but was less popular with loyal white citizens whose primary interest in fighting the Civil War was in defeating Southern secessionists and saving the Union.

Howe's genius in composing new lyrics was to equate abolishing slavery with both saving the country and righting a great moral wrong, thus appealing to a much larger audience.

Most Americans know the first verse, which starts "Mine eyes have seen the glory of the coming of the Lord/ He is trampling out the vintage where the grapes of wrath are stored" and repeats the chorus "Glory! Glory! Hallelujah!/His truth is marching on." Like many abolitionists, Howe was unapologetically religious, and, once the Civil War started, no pacifist. Her version of faith was known as the Church Militant, a Christian belief popular in the nineteenth century that the faithful were soldiers of God. The militant Christian's duty was to fight for righteousness on earth as a political activist, an elected official, or even as a uniformed volunteer shouldering a rifle.

In Howe's theology, God was on the side of abolition, and the U.S. Army had become a divine instrument to bring freedom to slaves. The song's fourth verse brings this point home vividly:

In the beauty of the lilies Christ was born across the sea,
With a glory in His bosom that transfigures you and me;
As He died to make men holy, let us die to make men free!
While God is marching on.

Wartime marching songs often go out of fashion after peace is declared. But "The Battle Hymn of the Republic" has continued to find people to sing it up to the present day. In a testament to its catchy tune and memorable lyrics, Howe's song has inspired numerous spinoffs and irreverent parodies ranging from Mark Twain's anti-imperialist "The Battle Hymn of the Republic, Updated" in 1901 to the labor-movement ballad "Solidarity Forever" to the schoolyard favorite "The Burning of the School" ("Mine eyes have seen the glory of the burning of the school/We have vanquished every teacher, we have broken every rule").

Today, there's no song which has gone that viral about saving the climate, installing solar panels, or humbling greedy oil executives. That standard of environmental protest events for half a century, Woody Guthrie's "This Land Is Your Land," is a powerful song that, through overuse, risks becoming an activist cliché. We need something just as singable today but more specific to climate disruption to rouse a new generation of citizens to action in what should be a very stirring fight.

Amanda Gorman, the Youth Poet Laureate who read "The Hill We Climb" at the inauguration of Joe Biden and Kamala Harris, could provide the lyrics if only some songwriter would put her words to a singable tune. Earlier, she wrote a poem starting from the iconic image of the Earth taken in 1968 on the Apollo 8 moon mission, "Earthrise," Dedicated to Al Gore, the poem is really about climate disruption: "Climate change is the single greatest challenge of our time."

This she offers as a challenge: "So I tell you this not to scare you/But to prepare you, to dare you /To dream a different reality."

Where despite disparities
We all care to protect this world,
This riddled blue marble, this little true marvel
To muster the verve and the nerve

To see how we can serve
Our planet. You don't need to be a politician
To make it your mission to conserve, to protect,
To preserve that one and only home
That is ours,
To use your unique power
To give next generations the planet they deserve.[52]

The Preacher: Rev. Lyman Beecher / Dr. James Hansen

Finally, we get to the preachers—and the climate scientists. It's not hard to see how these two specialized roles that seem to oppose each other across the chasm that separates faith from facts really share key characteristics in common. Scientists, as we discussed at the start of this chapter, were the first to sound the warning about climate disruption more than a century ago. Since then, the work of scientists has continued to be the basis of any good-faith discussion about climate.

Back in the nineteenth century, though ministers of the Gospel were not the most prominent leaders of abolition, most major American abolitionists, both Black and white, were Christians. Bible quotes and text drawn from the morality of Jesus permeated the speeches and writings of Frederick Douglass, William Lloyd Garrison, and other abolitionists.

The most famous abolitionist churchman was Rev. Henry Ward Beecher, brother of Harriet Beecher Stowe. Beecher pastored a Congregationalist church in New York City that catered to the city's business and cultural elite. Against Southern preachers who claimed that the Bible justified slavery, Beecher used his pulpit to preach a Christianity of abolition. Slavery and liberty were fundamentally incompatible, Beecher argued, making compromise impossible: "One or the other must die."

During the conflict over Kansas statehood, Beecher raised money to purchase Sharps rifles to send to abolitionist forces like the guerilla band of John Brown, saying that the guns would do more good than "a hundred Bibles," inspiring the press to nickname the crates of rifles "Beecher's Bibles."

No pacifist, Beecher strongly supported the Union war effort. After the leading Confederate army commanded by Robert E. Lee surrendered in April 1865, Abraham Lincoln sent Beecher down to the ceremony to reraise the U.S. flag at Fort Sumter, saying that "if it had not been for Beecher there would have been no flag to raise."

Since Beecher's time, religion and science have traded places. Now, science is the most credible arbiter of society's truth. So, though they rely on logic and data more than faith and inspiration, today's leading climate scientists have the same authority on climate today that Beecher had before the Civil War on slavery. James Hansen, the former NASA climatologist whose testimony to Congress in 1988 introduced climate science to the American public, is joined by other prominent climatologists who have gone beyond the laboratory to reach out to the public, including Michael Mann of "hockey stick" diagram fame and Katharine Hayhoe, an evangelical Christian with a special interest in reaching people of faith.

The Next Generation: Wide Awakes / Youth Activists

We've already discussed the Wide Awakes, an organization of young men who supported the election of Lincoln in 1860. After Fort Sumter, most members enlisted as a body in the U.S. Army to fight the slavocracy.

Today, the Sunrise Wide Awakes have adopted the same name as youth activists against slavery just before the Civil War, but many other young activists today work in the same spirit: Kelsey Juliana, plaintiff in a lawsuit of young people who sued the Trump administration for depriving them of their rights to future life, liberty and happiness; Haven Coleman, Isra Hirsi, and Alexandria Villaseñor, co-founders of U.S. Climate Strike; and Jamie Margolin, who at age 14 co-founded ZeroHour, a climate action group for young people.

For climate scientist Michael Mann, youth climate activists, combined with increasingly dramatic weather disasters, have led to a tipping point in climate activism.

"The Eye of Sauron is focused upon these kids. The most powerful industry in the world, the fossil fuel industry, sees them as an existential threat and has them firmly in its sights." Mann cites a recent meeting of OPEC, the international oil cartel, as proof. In July 2019, the group's secretary general, Mohammed Barkindo, called the youth climate movement "the greatest threat" the fossil fuel industry faces today. The OPEC chief was worried that political pressure from kids on oil companies was "beginning to...dictate policies and corporate decisions, including investment in the industry." Even the children of oil company execs are now asking their parents difficult questions about the future because "they see their peers on the streets campaigning against this industry."[53]

The Chorus: The Hutchinson Singers / Position Open

Formed by singer Jesse Hutchinson and his three brothers, this family choral group from New Hampshire featured nearly a dozen sons and daughters that toured the North before the Civil War, becoming the most popular entertainers nationwide in the 1840s. The group sang in four-part harmony a repertoire of political, social, comic, sentimental, and dramatic works. These included catchy tunes with memorable lyrics that managed to combine advocacy for social causes like temperance, rights for women and factory workers, and abolition with celebrations of exciting new technology of the day like the railroad.

A wonderful example is "Get Off the Track!" an abolitionist song written after members of the family choral group heard a lecture by William Lloyd Garrison. Its whimsical lyrics use the speed of train travel, whose speeds of 35 or 40 miles per hour were perceived as dizzyingly fast to people used to traveling on land by horse or mule or else on foot. The Hutchinsons used the figure of a moving train as an analogy for the unstoppable progress of the movement to emancipate slaves:

Ho! the Car Emancipation,
Rides majestic thro' our nation,
Bearing on its Train, the story,
Liberty, A Nation's Glory.
Roll it along, Roll it along, Roll it along,
thro' the Nation, Freedom's Car, Emancipation,
Roll it along, Roll it along,
Roll it along, thro' the Nation,
Freedom's Car, Emancipation.

Men of various predilections,
Frightened, run in all directions;
Merchants, Editors, Physicians,
Lawyers, Priests and Politicians.

Get out of the way! every station,
Clear the track of 'mancipation.

All true friends of Emancipation,
Haste to Freedom's Rail Road Station;
Quick into the Cars get seated,
All is ready and completed.
Put on the Steam! All are crying.
And the Liberty Flags are flying.[54]

In perhaps another casualty of today's fractured pop culture landscape and crowded musical market, there's no singer or musical group like the Hutchinsons today to get music lovers' feet tapping about fighting climate change. Is it even possible? It would be interesting to see somebody try.

Eminent Opponents of the Oil Power

Though dedicated to freedom and equality, the abolition movement was not shy about celebrating its leaders. Even when so much climate advocacy is of the "distributed" leaderless variety made possible by social media, the climate movement should also celebrate its leaders.

Political fights were more hierarchical in the past. Guided by egalitarian yearnings and enabled by the internet, today's activism is powered from the bottom up more than ever before. This is true on both the left and the right, whether you're fighting for civil rights, electric cars, or to ban abortions.

Sadly, it's also true that bottom-up activism inspired by deluded conspiracy theories like QAnon that have been doused in the gasoline of white supremacy also flourishes because the internet speeds the spread of lies as well as the truth. So, along with valuable groups like 350.org and the Sunrise Movement, the ease of coordinating a lot of independent activists spread all over the place has bolstered dangerous terrorist groups that participated in the U.S. Capitol insurrection of 2021 including the Oath Keepers, the Three Percenters, the Proud Boys, and various terrorist cells of neo-Nazis.

Distributed, do-it-yourself activism coordinated online is so appealing not only because it seems more democratic than traditional movements led by recognized leaders but also because social media activism uses new technology to do more with less.

But when you see how many ways there are for angry people to go wrong online, it's almost enough to make you nostalgic for the days of organized movements with famous leaders. At least such leaders could be held accountable. But more importantly, visible leaders provided models for their followers and inspiration for citizens who care about a cause to stay committed through the long years it takes to win. It's not misguided to want visible leaders in a cause that's up against powerful hierarchies and elites.

The climate movement is lucky to have leaders of conscience, courage, and conviction from Greta Thunberg to Al Gore to Alexandria Ocasio-Cortez to Bill McKibben to Varshini Prakash. People who want to save the climate should do a much better job of celebrating the valiant people who've stepped forward at great personal expense of time and treasure, enduring savage attacks from junkyard dogs for Big Oil and, in some cases, threats of physical violence. We need to celebrate our heroes for standing for what's right, just as abolitionists of old did.

The best argument against cynicism and defeatism—bigger threats to climate action than old-fashioned science denial—is someone who's willing to stand in front of a crowd and put hope into action with their own words and deeds. We need to recruit more people to win against the money and power of Big Oil. And people care much more about stories of other people than about stories of CO_2 or melting icebergs. Science may be the start of the climate battle, but people are what matters. The climate leaders we met here are just some of the tireless fighters for the rights of the oppressed, champions of truth and justice, and patriots in the best American tradition.

We need a poster depicting Eminent Opponents of the Fossil Fuel Power.

This is not hype. When you think about the gravity of the threat of climate breakdown, it's nothing less than the truth to celebrate those who fight for a livable future.

In the nineteenth century, abolition leaders helped millions of men, women, and children free themselves from the chains of slavery. Our leaders, if successful, will help us save our country and our civilization from climate chaos. As the past put up plaques to abolitionist liberators, so will future generations honor our climate heroes. Those heroes can inspire us today to make future history. By picturing our heroes today, we can add vigor to our fight, the kind of energy and commitment we will need to abolish fossil fuels and save our country and our world.

Conclusion: Winning Oil Abolition

We have to do with the past only as we can make it useful to the present and to the future. To all inspiring motives, to noble deeds which can be gained from the past, we are welcome. But now is the time, the important time. Your fathers have lived, died, and you must do your work.

—Frederick Douglass

This book has looked back at the past to inspire people in the present to save the world's climate to ensure humanity's future. As Frederick Douglass said, the past is only worth studying if we can make use of it today. The hope of this book is that people who want to abolish oil will find ideas and more importantly, inspiration, from the work of those who succeeded in abolishing slavery.

We have much in common with those abolitionists. We are all dreamers and doers aiming to right a wrong that has been allowed to stand for too long to the great peril of our country, our world, and each of ourselves. Some suffer more than others and the rest of us should acknowledge their suffering. Then we should do something about it, both to help them and to help ourselves.

Oil abolitionists today are ambitious because we have no choice. People who support the Green New Deal reject standalone solutions for climate in favor of overarching programs to build a clean economy while boosting economic and environmental justice. Likewise, abolitionists in the eighteenth and nineteenth century were ambitious and wanted to go beyond freeing slaves into making society more free, fair, and equal overall. Abolitionists didn't get everything they wanted, which led to wicked problems like persistent white supremacy that society is still grappling with today. But abolitionists did accomplish their main task, ending chattel slavery once and for all, while making many gains on racial equality.

We should pause a moment to reflect on what a heroic quest this proved to be. When the antislavery movement first started in the Age of Enlightenment, abolition seemed to many like an impossible dream. At the time, three out of four people worldwide lived under some system of bound labor, whether slavery, serfdom, peonage, indentured servitude, or even marriage where a wife gave up most of her property and personal rights to her husband. Slavery dated back thousands of years to the early days of civilization and was found across the globe. The ubiquity of slavery justified proslavery propagandists in claiming that the institution was an unavoidable and even beneficial part of human life, ordained by God and suited to the natural hierarchy of the races as affirmed by what qualified then as science.

When abolition succeeded in the nineteenth century, first in the British Empire in the 1830s and then in the United States in the 1860s, breaking the chains of millions of people was an accomplishment unprecedented in world history.

Freed people in the United States experienced their emancipation as a miracle of biblical proportions, "in religious and hysterical fervor," according to W.E.B. DuBois:

> This was the coming of the Lord. This was the fulfillment of prophecy and legend. It was the Golden Dawn, after chains of a thousand years. It was everything miraculous and perfect and promising. For the first time in their life, they could travel; they could see; they could change the dead level of their labor; they could talk to friends and sit at sundown and in moonlight, listening and imparting wonder-tales. They could hunt in the swamps, and fish in the rivers. And above all, they could stand up and assert themselves. They need not fear the [slave] patrol; they need not even cringe before a white face and touch their hats.[1]

Ending the institution of chattel slavery was monumental, but it wasn't enough for many abolitionists. They realized that without the political rights and economic opportunities taken for granted by white people, newly freed Black people would not be able to enjoy the benefits of freedom safely and securely. To secure those rights, during the post-Civil War period of Reconstruction, abolitionists went on to fight for the rights to vote and hold public office for men of all races (women would come later). They also fought for land reform and education necessary for emancipated people to support themselves independently, free of the control of plantation owners and overseers.

And, as if that wasn't enough, some more ambitious abolitionists in the United States hoped to use the opportunity of the disruptions to society and politics brought by the Civil War to enact progressive reforms that had been waiting for decades. These included votes for women, fair treatment for workers, and banning alcohol along with more personal improvements like promoting vegetarianism, making divorce easier for women to obtain, and propagating new religions.

For a century, historians sympathetic to the old Southern planter class presented Reconstruction as a failure. But historians today have concluded that the period got a bum rap and have started to revise our understanding of the time after the Civil War when the federal government tried to rebuild the devastated economies of the former Confederacy and integrate four million freedmen into the economy. Recent histories make clear that Reconstruction was really an ambitious attempt to lock in the gains of abolition for the future.[2]

The tragedy of this period came when the defeated but still defiant white planter elite of the former Confederate states refused to accept the verdict of the war and insisted on trying to roll back the clock to the days of slavery.

Reconstruction brought so many changes to politics and culture that historian Eric Foner has called it a second American revolution, albeit an unfinished one. W.E.B. DuBois called Reconstruction a massive experiment in multiracial democracy.

Freedom started with reuniting families. Freed from travel restrictions, formerly enslaved husbands went in search of wives and mothers sought children who had been sold off in the domestic slave trade. Freed people set up schools, sought economic independence by buying land when they could or otherwise seeking guarantees of fair pay and working conditions. They embraced their new rights as citizens. For the first time in the United States and perhaps anywhere, in the states of the former Confederacy occupied after the war by federal troops, Black and white leaders worked together to set up truly biracial governments. With Black men and poor white men alike now granted the vote for the first time, true democracy flourished.

Between 1870 and 1877, more than a dozen Black men were elected to Congress; 18 to state positions as lieutenant governors, treasurers, secretaries of state or superintendents of education; and at least 600 to state legislatures. Along with white allies, this new breed of leader representing the majority of the population of the South for the first time ever, came forward to pass visionary reforms including starting free public schools for Black and white students alike, integrating streetcars, and protecting workers and debtors.

In many cases Southern states under progressive Reconstruction governments passed reforms that reached so far that they surpassed policies in Northern states. Even though only five Northern states allowed Black men to vote at the time, in 1867 Radical Republicans in Congress required all states of the former Confederacy to pass new constitutions that allowed all Black men to vote. Ironically, the South, the poorest and most conservative section of the country, a region devastated by war, became a laboratory for innovations in democracy, economic opportunity, and racial justice.

In no other society that had liberated its slaves were freed people so quickly integrated into politics through the right to vote. After Haiti won its independence, the country was ruled by a series of military dictatorships for decades. In the British West Indies, after abolition in the 1830s, property requirements to vote remained so high that few former slaves qualified to cast ballots.

Only in the United States were all former slaves given full civil rights and then the right to vote by Congressional acts followed by Constitutional amendments. Even Radical Republicans in Congress who supported civil and voting rights for freedmen recognized that giving former slaves the vote only a few years after emancipation was unprecedented. "We have cut loose from the whole dead past," wrote Senator Timothy Howe of Wisconsin, "and have cast our anchor out a hundred years." [3]

This prophecy turned out to be correct. The Civil War, the most disruptive war in American history, was not enough to dislodge centuries of white supremacy. After that war, many white Americans, North and South, found civil and political rights for Black people to be a bridge too far. It would take another century, during the civil rights movement, to fully implement the agenda of Reconstruction. Even today, the fight for voting rights and equal treatment under the law continues.

The South has always been the nation's most conservative region, and even after Appomattox, its planter class remained resistant to change. In addition, the region was devastated by fighting and impoverished by the loss of many of its young men as well as its main source of wealth through abolition.

Prior to the passage of the Fifteenth Amendment that gave the vote to men across the nation regardless of race in 1870, most Northern states excluded Black men from the polls. Expecting the states of the former Confederacy to lead the nation in progressive change, to run at the vanguard of the revolution even ahead of progressive states in New England or out West, showed incredible idealism. But it also sowed the seeds of counterrevolution that led to Reconstruction's fall.

Under the watchful eye of armed federal troops, the old planter elite grudgingly accepted the failure of secession and the end of slavery, at least in name. But former slaveocrats were not willing to let go of either white supremacy or despotic control over their labor force. After the loss of their investment in slaves and whatever money they had in Confederate money and Southern government bonds that were now worthless, land was the only form of wealth that planters had left. To make that land pay on a payroll budget of near zero, planters needed to find farm workers who would pick cotton nearly for free. That meant reducing pay and working conditions for labor back down as close to slavery conditions as possible.

To achieve a workforce affordable enough for impoverished planters, workers had to be disempowered through rolling back the civil rights gains of Reconstruction, by law if possible and by force if necessary. White elites banded together in secret societies like the Ku Klux Klan or in open groups like the White League to stop the revolution in racial equality and take the South's social order back to a racial hierarchy as close to slavery as they could get away with. Through domestic terrorism, voter suppression, and electoral fraud, unreconstructed Confederates started to take back control, or "redeem," Southern states one by one even before the end of Reconstruction.

As a general, Ulysses S. Grant had led the U.S. Army to victory over the Confederacy in the Civil War, accepting Confederate commander Robert E. Lee's surrender at Appomattox. After the war, when Grant was elected president, he worked with Radical Republicans in Congress and former abolitionists including Frederick Douglass to protect Black civil rights. Grant directed the army to support Southern state governments still under biracial majority rule to fight back against the planter elite's attempted coups.

In 1871, Grant used federal troops to crush the Ku Klux Klan so successfully that the racist terrorist group didn't rear its head again for another half century. But, over time, the Northern voting public got tired of all the money and work it took to protect freedmen and their white allies in faraway Southern states.

After the contested presidential election of 1876, leaders of the Republican and Democratic parties agreed to a compromise to let the Republican Rutherford Hayes enter the White House if he would commit to remove federal troops from Southern statehouses and give a free hand to white elites to take back control of the state governments of the South.

When Hayes did send federal troops back to their barracks the following year, it spelled the effective end of Reconstruction. By 1877, the former Slave Power regained power in every Southern state, though sometimes biracial coalitions managed to wrest back control for a brief period, as in Virginia in the 1880s or North Carolina in the 1890s.

All-white "redeemer" governments dominated by former Confederates set up apartheid regimes that amounted to police states from Virginia to Texas. Redeemers suppressed voting and participation in government by poor whites and Black men and then institutionalized white supremacy through Jim Crow segregation. New laws allowing mass incarceration with prison labor and other exploitative employment practices returned many Black farmworkers to a form of bound labor that was all too similar to slavery. Reconstruction was a noble revolution, but one that was forced backward.

"The slave went free; stood a brief moment in the sun; then moved back again toward slavery," wrote W.E.B. DuBois in the 1930s.[4]

America would be a better place today if the voting public and the business leaders of the North had stood behind Reconstruction and prevented the old Slave Power from overthrowing biracial democracy in Southern states.

But not everything that Reconstruction accomplished was destroyed. The two cornerstones of Black life in America, the Black family, never safe under slavery, and the Black church, remained intact. Public schools were firmly established as a responsibility of state government for students of all races. Despite oppressive labor laws, conditions for most farmworkers and tradesmen were a big improvement over slavery. The doors of economic opportunity opened by Reconstruction were not entirely closed, as shown by the emergence of a small but growing group of Black landowners, businesspeople, and professionals. And though rarely in practice, but at least in law, Black people both South and North were now citizens with civil rights that had to be respected.

All these elements would prepare the country for the advances of the civil rights movement in the 1950s and 1960s, sometimes called the Second Reconstruction. "Perhaps the remarkable thing about Reconstruction was not that it failed, but that it was attempted at all and survived as long as it did," writes historian Eric Foner.[5]

The rise and fall of Reconstruction offers lessons to the climate movement, mainly that progress does not always go one way and sometimes it can be reversed. Also, if a hated industry decides to fight back with lies and misinformation, and if they spend enough money to spin a myth and keep developing it for years, then they can win even after losing.

We've seen how in the 1980s, after early efforts by Al Gore to bring climate change to public attention, fossil fuel companies launched a decades-long propaganda campaign to deny climate science and protect themselves from government restrictions on producing and selling dirty energy. A century earlier, to maintain white control of Southern states without federal interference, the planter elite ran their own national propaganda campaign against civil rights action centering on the myth of the noble "Lost Cause" of the Old South.

This myth painted the slave states of the antebellum era as a happy land of kind masters and contented slaves, portrayed the Civil War as a tragic conflict between brave white men on both sides artificially whipped up by misguided abolitionists, and depicted Reconstruction as the wholesale rape of Southern society by ignorant and vengeful freed slaves egged on by a self-serving alliance of local white Republicans, known as "scalawags," and "carpetbaggers," Northern Republicans who came South.

One of the most successful examples of history being written not by the proverbial winners of a war but instead by its losers, the Lost Cause myth conquered vast territories of the American mind and popular culture. Leading histories of the war and Reconstruction, bestselling novels of plantation life, popular live musical theater performances known as "minstrel shows," and two of the most important Hollywood films of all time, *The Birth of a Nation* and *Gone with The Wind*, all reflected a point of view hostile to abolition and Black equality and sympathetic to slavery, white supremacy, and the rule of the white planter class.

Lost Cause propaganda accomplished its goal and helped convince white Americans to let Southern planters "solve" racial problems on their own without interference from the federal government. Southern segregation and white supremacist rule would not be seriously challenged until the civil rights movement, nearly a century after the Thirteenth Amendment ended slavery. Even today, the work of Reconstruction across the South and across the nation remains unfinished, as discussions about America's racial disparities after the death of George Floyd reminded us.

Despite their firm commitment to civil rights and equality, abolitionists failed in ending racial discrimination and economic exploitation of Black Americans. And abolitionists certainly didn't get to implement the many progressive reforms conceived by some of the most wide-open minds, most compassionate hearts, and stiffest backbones in American history.

The abolition revolution remained an unfinished one. But it was still a revolution and one that rocked America and the world. Four million Black Americans were freed from chains and the federal government did an about-face, transforming from a protector of slavery to a guarantor of freedom, no matter how imperfect its execution. And by laying the framework for a society based not on white supremacy but on racial equality, abolition made possible the civil rights movement of the 1950s and 1960s and all other movements for racial fairness since then from Black Power to Black Lives Matter.

Abolition and Reconstruction continue to retain the power to inspire political change. Rev. William Barber, founder of the Moral Mondays movement and the Poor People's Campaign, has called for a new Reconstruction movement today that sounds a lot like the Green New Deal. His idea for a program to promote both civil rights and economic equity goes beyond dry discussions of public policy to draw on deep reserves of moral fervor reminiscent of the first Reconstruction or the abolition movement.

"Moral language can re-frame and critique public policy regardless of who's in power," Barber writes. "A moral movement claims higher ground than merely a partisan debate, something that's bigger than left versus right, conservative versus liberal. We have to begin to re-frame the conversation not to talk about left policies and right policies, but let's talk about violence."[6]

As we've seen, abolition also created a model of political movement driven by citizens rather than political parties or special interests. In building public awareness and support to drive action by presidents and governors, legislatures and courts, regulators and administrators, abolition bequeathed powerful tools to causes that followed, from women's suffrage to Occupy Wall Street. And in achieving victory over the most powerful moneyed interest of its day—the Slave Power of cotton planters and their allies in manufacturing, shipping, and finance—abolition offers the best model for people who want to save the climate to triumph over the single most powerful force destroying the climate today, fossil fuel companies.

It took nearly a century to abolish slavery on both sides of the Atlantic. We don't have that much time to save the climate and abolish oil worldwide, starting with the dark heart of the global Oil Power in the United States. Fortunately, it will be easier to end oil now than it was to end slavery back then. Abolishing oil is the challenge of our lifetimes.

But this epic fight, though an issue of life and death, will be easier to win than abolishing slavery, which was bound up with a social and political system based on white supremacy that still poisons American life today. Fortunately, for most people fossil fuels are not integral to their personal identity or their place in society. It will be much easier to replace oil, gas, and coal with solar and wind power than it was to free four million slaves and then accustom white people to treating them as equals.

The challenge in replacing dirty energy with clean energy is not social but economic. And that makes it much easier to address. We do need to make transformative changes to the economy. But we don't need to overturn a whole social and political system, though we do need to address the effects of old racial hierarchies. We don't even need to abolish capitalism.

Whether we need to stop and reverse economic growth or can simply replace dirty growth with clean growth, probably with a pace that's slower and a location that's shifted from the richest nations to poorer countries in the global South, is still up for debate. But either way, a more purposeful economy designed to remain within the limits of the earth's resources can still provide us with modern comforts and prosperous lives.

Today it seems obvious that manufacturing and automation would take the place of slave labor for economic reasons alone. But before the Civil War, that wasn't obvious to either slaveowners or abolitionists. American slavery was not shrinking but growing during the first half of the nineteenth century. Slave agriculture reached its height in 1860, the very year before the Civil War started. Without that war and the Thirteenth Amendment the war made possible, it's likely that slavery would have continued for decades longer in the United States. Until late in the war even Abraham Lincoln thought slavery couldn't be completely abolished until the year 1900.

Just before its end, slavery was riding high. Fortunately, that's not the case today with fossil fuels today. Oil, gas, and coal companies have been shrinking for years. Smaller drillers and mining operations have disappeared into bankruptcy or been acquired at fire sale prices by bigger and more solvent competitors.

Yet, for all their acquisitions, the remaining companies are less profitable and less powerful than they've been in decades. Meanwhile, everybody knows that the alternative to fossil fuels is readily available in the form of solar and wind power backed up by batteries to store energy, all technologies that have gotten so affordable and grown so fast that they've entered the mainstream. The question is no longer whether clean energy will displace dirty energy but simply how soon it will happen.

Dozens of major plans developed by experts in technology and business are circulating to transition the global economy to 100 percent clean energy by 2050 or earlier. To just take one out of the many reports that have been published in the last few years, researchers at Stanford University project that it's both doable and affordable to move the global economy to entirely clean energy by 2050. The upfront cost of $73 trillion would be repaid in seven years and would create 28.6 million more full time jobs than if countries continue using fossil fuels.[7]

But spending money won't be enough. "When it comes to cutting carbon, the stick may be just as important as the carrot—perhaps more so," writes Elizabeth Kolbert. "Putting up wind turbines doesn't, in itself, accomplish much for the climate: emissions fall only when fossil-fuel plants are shuttered."

The Biden administration seems to understand that fighting climate is not all about addition but also will require some subtraction—as clean energy is added, dirty energy must also be taken away. Biden's initial plan not only proposed cutting fossil fuel subsidies worth up to $50 billion a year but also would require electric utilities to produce a portion of their power from renewable sources. Connecting these cuts to creating jobs may make climate action more palatable to politicians, but friends of the Oil Power will still object to shutting down drilling operations, refineries, and pipelines.

"Unfortunately, though, the laws of geophysics are indifferent to politics," as Kolbert puts it.[8]

Genevieve Guenther, the founder of End Climate Silence, an organization that works to promote accurate media coverage of the climate crisis, was impressed with Biden's first few months in office. So far, Biden has been the best president ever on climate action. But Guenther was afraid that the fossil fuel industry would go on the offensive and that the administration and the climate movement wouldn't be ready.

Just like the Slave Power of old, the Oil Power would use clever arguments order to confuse the public and defer serious action. Because these arguments always contain a small grain of truth, they're especially hard to spot as propaganda.

"People can recognize fossil fuel industry talking points by thinking about what they're designed to do. In general, fossil fuel talking points are designed to do three things: make people believe that climate action will hurt them and hurt their pocketbooks in particular; make people think we need fossil fuels; and try to convince us that climate change isn't such a big deal."[9]

The answers to these talking points are easy for anybody even a little familiar with the Green New Deal. First, replacing fossil fuels with clean energy will create more jobs than are lost. And the program will focus on retraining people who work in dirty energy to work in clean industries.

Of course, the transition won't be painless for all workers, and that's why the Green New Deal calls for a stronger social safety net of healthcare for all and better social welfare programs. Numerous plans have outlined how all fossil fuels can be replaced by either using less energy or getting the energy that's still needed from solar and wind power and other clean sources.

Then, most climate scientists, combined with most leaders of world governments, corporations, and churches agree that climate chaos is a major threat to world order and the future of the human species. Anyone who denies or downplays the dangers of our heating atmosphere is just spreading propaganda and confusion.

Guenther isn't worried that fossil fuel companies are correct. Quite the opposite: she knows that they're lying. But she's concerned that many voters may be fooled and that elected officials won't be able to withstand pressure from Big Oil propaganda and lobbying. Leaders from the president on down must be explicit about the threat of climate disruption and the purpose for any solutions they support. "I worry that the Biden administration isn't bringing that message to the foreground, because you need that to be part of the understanding of why we're doing this work."

In the face of fear and doubt about the rightness of our cause, Guenther recommends courage and conviction. "The motivation here is that we're trying to save our world. We're trying to save the lives of our children. I think activists do a pretty good job of keeping that messaging in the foreground, but I really wish that politicians would do it too. I think they're still running scared, and I don't think they have to be." You could say she's just asking climate people to take a page from the abolitionists of old, to stand bold in the face of attacks.

Guenther's organization, End Climate Silence, argues that the news media is key to protecting leaders who want to do the right thing from propaganda and lobbying attacks by the Oil Power. Like so many big things in politics today, Guenther's effort started with a Tweet.

In 2018, Guenther complained on Twitter how three NPR stories that were clearly about climate change—droughts in the Pacific Northwest, flooding in Japan, and self-driving cars—didn't mention climate. She got a reply from Chris Hayes of MSNBC.

This is the same Chris Hayes who, we saw near the beginning of this book, published an article several years earlier drawing an extended analogy between slavery abolition and fighting climate chaos. Hayes obviously cared deeply about climate. Yet, in clearly in frustration, Hayes tweeted, about his show, which is very progressive since it's on MSNBC, "every single time we've covered climate change, it's been a palpable ratings killer."

Guenther replied with some advice: Instead of doing discrete stories about climate change, why don't you just mention climate change and its effects in the stories you're already reporting? Hayes tweeted back that he could do that. This gave birth to Guenther's campaign "to pressure journalists and news anchors to mention climate change in the stories they're already reporting about its effects."[10]

Journalists and producers must stop segregating climate change in environmental news, Guenther says. As a political concern, the environment is a low priority for most Americans compared to the economy, national security, and social issues. "Climate change is not just a topic for the science or environment section. It's the essential context for stories about extreme weather, energy, politics, business and finance, immigration, real estate, travel, health, food, sports, and the arts."

Fair and accurate journalism requires reporters and editors to make the connection to climate in stories whenever that connection is there, which is much more often than journalists seem to think. As her group's website puts it:

Heat waves. droughts. fires. floods. rain bombs. hurricanes. typhoons. crop failure. water scarcity. insect-borne disease. refugee crises. financial losses. geopolitical instability. These stories are all around us. The media reports on them every day. But they fail to connect what's happening to climate change.[11]

End Climate Silence urges readers and viewers to alert the group to news stories that should have talked about climate but didn't.

In the Internet Age, the news media is still important, but they're no longer the only game in town. When anybody can become a publisher through their social media accounts, ordinary citizens also have an important role to play on their own in telling a more accurate and more impactful story about climate. So, citizens should spread the word on their own even if this takes them outside their comfort zone, says Guenther: "Talk about climate breakdown with your friends, co-workers, and acquaintances, especially when it seems most awkward to do so."[12]

Dedicating your personal life to climate evangelism and to abolishing oil is squarely in the tradition of abolitionists. It will take time to figure out the right mix of solutions to save the climate. Though morally distasteful, solutions may include some version of compensation for owners, to neutralize their opposition. Climate solutions must include reparations to frontline communities, for justice and to spur economic development.

Religious faith was important to abolitionists and Christianity has remained important to champions of racial justice across the world. Jim Wallis, founder of the progressive Christian group Sojourners, writes about a time that he witnessed former South African archbishop Desmond Tutu stand up to government intimidation during the fight against apartheid.

The Black religious leader was preaching a sermon at a service at St. George's Cathedral in Cape Town, now renowned as the "People's

Cathedral" for its leading role in the resistance to apartheid. While Tutu was speaking, members of the state security police started to enter the church.

Holding tape recorders and scribbling on note pads, the agents were gathering evidence of criticisms against government policy by Tutu that would violate the repressive regime's restrictions on freedom of speech. The implied threat was not idle—Tutu and other church officials had been arrested just a couple weeks earlier and held in jail for several days for outspoken remarks against apartheid.

The authorities in Pretoria wanted to demonstrate that religious leaders who criticized the apartheid regime would be treated just like any other government critics. Would the authorities arrest Tutu now to show that wearing a clerical collar provided no promise of more lenient treatment?

Wallis describes how Tutu responded from the pulpit when he saw the government agents:

> After meeting their eyes with his in a steely gaze, the church leader acknowledged their power ("You are powerful, very powerful") but reminded them that he served a higher power greater than their political authority ("But I serve a God who cannot be mocked!"). Then, in the most extraordinary challenge to political tyranny I have ever witnessed, Archbishop Desmond Tutu told the representatives of South African apartheid, "Since you have already lost, I invite you today to come and join the winning side!"

Though he smiled when he said it, Tutu's friendly invitation was also a clear act of defiance that was noted by the congregation. "The crowd was literally transformed by the bishop's challenge to power. From a cowering fear of the heavily armed security forces that surrounded the cathedral and greatly outnumbered the band of worshipers, we literally leaped to our feet, shouted the praises of God and began...dancing."

Wallis and the rest of the congregation left the service dancing, spilling out through the cathedral's entrance, over its steps, and into the plaza beyond. Police and soldiers waiting outside did not plan for a confrontation with dancing worshippers. Taken by surprise, instead of making arrests, the lines of uniformed officers backed up to give the crowd more space to whirl around and jump for joy.[13]

This story offers a powerful lesson for anyone who wants to save the climate by abolishing oil.

People who promote fossil fuels may still have immense power, just as leaders of the apartheid regime in South Africa did in the 1980s or leaders of the Slave Power did in the United States in the 1860s. But we who labor to save the climate serve powers that cannot be mocked, whether nature or God. Anyone who works for the Oil Power has already lost. So, let's invite them to join the winning side!

What would that mean?

First, here's what it would not mean. Saving the climate would not mean just deciding to believe advertising from oil companies that they want to be part of the solution on climate. And it would certainly not mean naively accepting plans by oil companies to supposedly achieve "net-zero" carbon emissions decades in the future even while they continue to produce fossil fuels for decades to come. Dirty energy corporations can never plant and maintain enough trees in the Amazon to suck up the pollution they still plan to keep pumping out by continuing to produce oil, gas, and coal.

We don't need any more "innovation" in repurposing fossil fuels into allegedly green products. We don't need more clever ways to turn petroleum into plastics or natural gas into hydrogen or coal into paperweights.

What saving the climate requires is simple: abolishing fossil energy. It's about really quitting fossil fuels. It's about leaving almost all the remaining oil, coal, and gas in the ground. It's about shutting down the dirty energy industry forever. It may even mean speeding up the process by directing the government to buy out fossil fuel companies, sell or destroy their assets, and then shut the companies down permanently.

As we dance our way to a world beyond oil, we know what winning looks like.

Winning looks like a stable climate with weather as much as possible like what humans have gotten used to since we spread out to all four corners of our mostly hospitable globe.

Winning looks like heat, light, and bytes from clean renewable energy affordable enough for anybody of any race, gender, or income who wants them anywhere.

Winning looks like freedom instead of slavery, peace instead of war, democracy instead of dictatorship, quality instead of quantity, and no need to make a choice between people and profits.

Winning looks like racial equity, clean communities for all regardless of race or wealth, clean-up for past pollution and reparations for past injustices, and good jobs for all who want them.

Winning looks like some personal lifestyle changes by citizens and consumers and many more changes by governments in laws, regulations, and taxes to make the world's economies clean, fair, and prosperous.

Winning looks like breaking the chains that bind all people to oil while welcoming all people to help us break the chains of the Oil Power.

And to the people of the fossil fuel industry, whether you work in drilling or refining, in lawyering or lobbying, we invite you today to come join the winning side.

Acknowledgements

My wife Lindsay provided love and support, including many discussions and many readings of drafts. Mike Shelton also provided support through discussions and reading the manuscript, as did Joy Loving and Nelson Patterson. Steve Corneliussen suggested the book's subtitle and gave much good advice. All errors and omissions are my own.

Notes

Introduction

[1] Emily Sanders, "Big Oil CEOs Weren't at the Senate Budget Committee's Climate Hearing, but the Industry Made its Mark," *ExxonKnews*, April 16, 2021, https://exxonknews.substack.com/p/big-oil-ceos-werent-at-the-senate.

[2] Following the practice of *Columbia Journalism Review*, this book capitalizes "Black," when talking about people, even in quotes where it was not originally capitalized, but does not capitalize "white." "For many people, *Black* reflects a shared sense of identity and community. White carries a different set of meanings; capitalizing the word in this context risks following the lead of white supremacists." See https://www.cjr.org/analysis/capital-b-black-styleguide.php. Since "brown" doesn't carry the same importance to define a community, this book doesn't capitalize that term either when talking about people.

[3] Richard Collett-White, "David Attenborough: Climate Change May Become Abhorred as Much as Slavery," *Climate Home News*, September 7, 2019, https://www.climatechangenews.com/2019/07/09/david-attenborough-climate-change-may-one-day-abhorred-like-slavery/.

[4] In a poll conducted just after the November 2020 elections by Yale and George Mason Universities, 66% of Americans said that developing sources of clean energy should be a high or very high priority. See John Schwartz, "Survey Finds Majority of Voters Support Initiatives to Fight Climate Change," *New York Times*, January 15, 2021, https://www.nytimes.com/2021/01/15/climate/climate-change-survey.html.

[5] Damian Carrington, "Al Gore: Battle Against Climate Change Is Like Fight Against Slavery," *The Guardian*, June 21, 2017, https://www.theguardian.com/environment/2017/jun/21/al-gore-battle-against-climate-change-like-fight-against-slavery.

[6] Naomi Klein, *This Changes Everything: Capitalism vs. the Climate* (New York: Simon & Schuster, 2014), 255.

[7] Chris Hayes, "The New Abolitionism: Averting Planetary Disaster Will Mean Forcing Fossil Fuel Companies to Give Up at Least $10 Trillion in Wealth," *The Nation*, April 22, 2014, https://www.thenation.com/article/archive/new-abolitionism/.

[8] Denise Fairchild, "What Can the Abolitionists Teach Us About Climate Change?" *Urban Resilience Project*, December 22, 2016, https://medium.com/@UrbanResilience/what-can-the-abolitionists-teach-us-about-climate-change-e34765b88c81.

[9] Peter Sinclair, "Berkeley Study: 90% Carbon-Free Electricity Achievable by 2035," *Yale Climate Connections*, September 11, 2020, https://yaleclimateconnections.org/2020/09/video-berkeley-study-90-percent-carbon-free-electricity-achievable-by-2035/.

[10] Breeanna Hare and Doug Criss, "Six Questions about Slavery Reparations, Answered," *CNN*, August 15, 2020, https://www.cnn.com/2020/08/15/us/slavery-reparations-explanation-trnd/index.html.

[11] Sam Evans-Brown, "Windfall Part 1: Sea Change," *OUTSIDE/IN,*" June 24, 2021, https://outsideinradio.org/transcript-windfall-part-1-sea-change.

[12] Library of Congress, "Women Authors," *From Slavery to Freedom: The African-American Pamphlet Collection, 1822-1909*, http://memory.loc.gov:8081/ammem/collections/pamphlets/aapcpres07.html.

[13] Lewis Freeman, interview with the author, August 12, 2020.

[14] Black and Hispanic/Latino Americans are more likely be alarmed or concerned about global warming than whites, according to a 2019 survey. See Matthew Ballew *etal.*, "Which Racial/Ethnic Groups Care Most about Climate Change?" Yale Program on Climate Change Communication, April 16, 2020, https://climatecommunication.yale.edu/publications/race-and-climate-change/.

[15] Kendra Pierre-Louis, "Understanding the Fossil Fuel Industry's Legacy of White Supremacy," *DeSmog*, Mar 29, 2021, https://www.desmogblog.com/2021/03/29/fossil-fuel-industry-legacy-white-supremacy.

[16] Bill McKibben, "Racism, Police Violence, and the Climate Are Not Separate Issues," *The New Yorker*, June 4, 2020, https://www.newyorker.com/news/annals-of-a-warming-planet/racism-police-violence-and-the-climate-are-not-separate-issues.

[17] William J. Barber, "Rev. Dr. William J. Barber II on: The Climate Crisis and Social Justice," Climate Reality, July 20, 2020, https://www.youtube.com/watch?v=jsEEEyC8Oms.

Chapter 1

[1] "This Professor Wants You to Give Up Your Climate Guilt," *Grist*, April 8, 2020, https://grist.org/fix/this-professor-wants-you-to-give-up-your-climate-guilt/.

[2] James Brewer Stewart, "From Moral Suasion to Political Confrontation: American Abolitionists and the Problem of Resistance 1831-1861," *Abolitionist Politics and the Coming of the Civil War* (Amherst, MA: University of Massachusetts Press, 2008), 3-32.

[3] James Brewer Stewart, "Is the Green New Deal Impossible?" *History News Network*, February 14, 2019, https://historynewsnetwork.org/article/171258.

[4] Kate Aronoff, "We Didn't Start the Fire," *Winning the Green New Deal: Why We Must, How We Can*, edited by Varshini Prakash and Guido Girgenti (New York: Simon & Schuster, 2020), 25.

[5] Charles Duhigg, "Clean Water Laws Are Neglected, at a Cost in Suffering," *New York Times*, September 12, 2009, https://www.nytimes.com/2009/09/13/us/13water.html.

[6] Quoted in Nathanael Johnson, "Have You Been Doing Environmentalism Wrong?" *Grist*, April 28, 2021, https://grist.org/climate-energy/have-you-been-doing-environmentalism-wrong/.

[7] Brian Eckhouse, "World Added More Solar, Wind Than Anything Else Last Year," *Bloomberg Green*, September 1, 2020,

https://www.bloomberg.com/news/articles/2020-09-01/the-world-added-more-solar-wind-than-anything-else-last-year.

[8] Daisy Nguyen, "Approvals for New Oil and Gas Wells up in California," *Associated Press*, September 20, 2020, https://apnews.com/article/d04910d29539d39e24eaa725bcf4545f.

[9] Rachel Frazin, "Biden Admin Backs Trump-Era Approval of Controversial Line 3 Pipeline Permit," *The Hill*, June 24, 2021, https://thehill.com/policy/energy-environment/560037-biden-administration-backs-trump-era-approval-of-controversial-line.

[10] Robert Brulle, "The Structure of Obstruction: Understanding Opposition to Climate Change Action in the United States," Climate Social Science Network, Brown University, April 2021, https://www.cssn.org/wp-content/uploads/2021/04/CSSN-Briefing_-Obstruction-2.pdf.

[11] Holly Jackson, *American Radicals: How Nineteenth Century Protest Shaped the Nation* (New York: Random House, 2019), xiii.

[12] Holly Jackson, *American Radicals*, 89.

[13] Holly Jackson, *American Radicals*, 94.

[14] Chris Hayes, "The New Abolitionism."

[15] Naomi Klein, *This Changes Everything: Capitalism vs Climate* (New York: Simon & Schuster, 2014), 455.

[16] Heather Long and Andrew Van Dam, "West Virginia's Surprising Boom, and Bust, Tells The Story of Trump's Promise to Help The 'Forgotten Man'," *Washington Post*, October 30, 2020, https://www.washingtonpost.com/business/2020/10/30/trump-economy-west-virginia-forgotten-man/.

[17] Sarah Hsu, "When Band-Aids won't Cut it: The Climate Crisis is a Health Crisis," *World War Zero Magazine*, September 20, 2020, https://worldwarzero.com/magazine/2020/09/bandaids-wont-cut-it-climate-crisis-health-crisis/.

[18] Emily Holden, "Revealed: How The Gas Industry is Waging War Against Climate Action," *The Guardian*, August 20, 2020, https://www.theguardian.com/environment/2020/aug/20/gas-industry-waging-war-against-climate-action.

[19] Sarah Ponczek and Katherine Greifeld, "Exxon Booted from Dow Industrials in Major Embrace of Tech," *BloombergQuint*, August 25, 2020, https://www.bloombergquint.com/global-economics/dow-industrials-kicks-out-exxon-in-biggest-shakeup-since-2013.

[20] Antonia Juhasz, "The End of Oil Is Near: The Pandemic May Send The Petroleum Industry to The Grave," *Sierra*, Aug 24 2020, https://www.sierraclub.org/sierra/2020-5-september-october/feature/end-oil-near.

[21] Reuters, "Energy Executives Say US Oil Production Has Peaked: Dallas Fed Survey," *Energy World*, September 24, 2020, https://energy.economictimes.indiatimes.com/news/oil-and-gas/energy-executives-say-us-oil-production-has-peaked-dallas-fed-survey/78287045.

[22] Eugene R. Dattel, "Cotton in a Global Economy: Mississippi (1800-1860)," *Mississippi History Now,* http://mshistorynow.mdah.state.ms.us/articles/161/cotton-in-a-global-economy-mississippi-1800-1860.

[23] Kevin Crowley and Akshat Rathi, "Exxon's Plan for Surging Carbon Emissions Revealed in Leaked Documents," *Bloomberg Green,* October 5, 2020, https://www.bloomberg.com/news/articles/2020-10-05/exxon-carbon-emissions-and-climate-leaked-plans-reveal-rising-co2-output.

[24] Antonia Juhasz, "Bailout: Billions of Dollars of Federal COVID-19 Relief Money Flow to the Oil Industry," *Sierra,* Aug 26, 2020, https://www.sierraclub.org/sierra/bailout-billions-dollars-federal-covid-19-relief-money-flow-oil-industry.

[25] Senate Democrats' Special Committee on the Climate Crisis, *The Case for Climate Action: Building A Clean Economy for The American People,* August 25, 2020, 199.

[26] Senate Democrats' Special Committee on the Climate Crisis, *The Case for Climate Action,* 200.

[27] Hiroko Tabuchi, Michael Corkery, and Carlos Mureithi, "Big Oil Is in Trouble. Its Plan: Flood Africa with Plastic," *New York Times,* August 30, 2020, https://www.nytimes.com/2020/08/30/climate/oil-kenya-africa-plastics-trade.html.

[28] Gabriella Paiella, "How Would Prison Abolition Actually Work?" *GQ,* June 11, 2020, https://www.gq.com/story/what-is-prison-abolition.

[29] Robert Bullard, "Time for Whites to Stop Dumping Their Pollution on People of Color," DrRobertBullard.com, April 2, 2019, https://drrobertbullard.com/time-for-whites-to-stop-dumping-their-pollution-on-people-of-color/.

[30] Bill McKibben, "How We Got to the Green New Deal," *Winning the Green New Deal,* 64.

[31] Bill McKibben, "Racism, Police Violence, and the Climate Are Not Separate Issues."

[32] Matthew Ballew *et al.,* "Which Racial/Ethnic Groups Care Most About Climate Change?"

[33] Ian Haney López, "Averting Climate Collapse Requires Confronting Racism," *Winning the Green New Deal,* 50.

[34] https://howmanytimeshasthehousevotedtorepealobamacare.com/.

[35] Jim Percoco, "The United States Colored Troops," American Battlefield Trust, https://www.battlefields.org/learn/articles/united-states-colored-troops.

[36] John Parkinson, "Rep. Alexandria Ocasio-Cortez, Expanded 'Squad' Demand Biden Deliver on Green New Deal," *ABC News,* November 20, 2020, https://abcnews.go.com/Politics/rep-alexandria-ocasio-cortez-expanded-squad-demand-biden/story?id=74297506.

[37] Andrew Marantz, "Are We Entering a New Political Era?" *The New Yorker,* May 24, 2021, https://www.newyorker.com/magazine/2021/05/31/are-we-entering-a-new-political-era.

[38] Damon Centola *et al.,* "Experimental Evidence for Tipping Points in Social Convention," *Science,* June 8, 2018, https://science.sciencemag.org/content/360/6393/1116.

[39] Eric Foner, "The Mexican War & Expansion of Slavery," *MOOC: The Civil War and Reconstruction, 1850-1861,* Columbia University, October 27, 2014, https://youtu.be/KkgkZwQ9HgQ.

Chapter 2

[1] Quoted in Harriet Beecher Stowe, *The Annotated Uncle Tom's Cabin*, edited by Henry Louis Gates Jr. and Hollis Robbins (New York: W.W. Norton, 2006), 378.

[2] George Fitzhugh, "Southern Thought," *The Ideology of Slavery: Proslavery Thought in the Antebellum South, 1830-1860*, edited by Drew Gilpin Faust (Baton Rouge, LA: LSU Press, 1981), 289.

[3] George Fitzhugh, "Southern Thought," 279.

[4] Peter Sinclair, "Berkeley Study: 90% Carbon-Free Electricity Achievable by 2035," *Yale Climate Connections*, September 11, 2020, https://yaleclimateconnections.org/2020/09/video-berkeley-study-90-percent-carbon-free-electricity-achievable-by-2035/.

[5] Eleanor Barkhorn, "'Vote No on Women's Suffrage': Bizarre Reasons For Not Letting Women Vote," *The Atlantic*, September 6, 2012, https://www.theatlantic.com/sexes/archive/2012/11/vote-no-on-womens-suffrage-bizarre-reasons-for-not-letting-women-vote/264639/.

[6] https://www.equalrightsamendment.org/era-ratification-map.

[7] Nathaniel Rich, *Losing Earth: A Recent History* (New York: MCD Books, 2019).

[8] See especially the 1996 best-selling book *The Fourth Turning: What the Cycles of History Tell Us About America's Next Rendezvous with Destiny* (New York: Crown, 1997) by Neil Howe and William Strauss, which also divides American history into eighty-year cycles, each starting and ending with a national crisis, just as Friedman does. And while Friedman only goes back 250 years to the American Revolution, Strauss and Howe go back to the fifteenth century to track 500 years of eighty-year cycles in Anglo-American history, adding more support for the validity of this approach. Historians who analyze history in terms of repeating cycles include Arthur Schlesinger, Jr. in *The Cycles of American History* (Boston: Houghton Mifflin Company, 1986) and Peter Turchin in *War and Peace and War* (New York: Plume, 2007).

[9] George Friedman, *The Storm Before the Calm: America's Discord, the Coming Crisis of the 2020s, and the Triumph Beyond* (New York: Doubleday, 2020), 150.

[10] George Friedman, *The Storm Before the Calm*, 187, 189.

[11] Kate Raworth, "A Healthy Economy Should be Designed to Thrive, Not Grow," TED2018, April 2018, https://www.ted.com/talks/kate_raworth_a_healthy_economy_should_be_designed_to_thrive_not_grow/.

[12] See Edward E. Baptist, *The Half Has Never Been Told: Slavery and the Making of American Capitalism* (New York: Basic Books, 2014).

[13] David Blight, *The Civil War and Reconstruction Era, 1845-1877*, Open Yale Courses, Yale University, Spring 2008, https://oyc.yale.edu/history/hist-119.

[14] https://www.in2013dollars.com/us/inflation/1860?amount=1.

[15] Ben Winters, *Underground Airlines* (New York: Mulholland Books, 2017).

[16] Kate Hodal, "One in 200 People Is a Slave. Why?" *The Guardian*, February 25, 2019, https://www.theguardian.com/news/2019/feb/25/modern-slavery-trafficking-persons-one-in-200.

Chapter 3

[1] Michael Mann, *The New Climate War*, 230.

[2] David Michaels, *Doubt is Their Product: How Industry's Assault on Science Threatens Your Health* (New York: Oxford University Press, 2008), 3.

[3] Benjamin Franta, "On its 100th Birthday In 1959, Edward Teller Warned The Oil Industry About Global Warming," *The Guardian*, January 1, 2018, https://www.theguardian.com/environment/climate-consensus-97-percent/2018/jan/01/on-its-hundredth-birthday-in-1959-edward-teller-warned-the-oil-industry-about-global-warming.

[4] Matthew Taylor, "Climate Emergency: What The Oil, Coal and Gas Giants Say," *The Guardian*, October 10, 2019, https://www.theguardian.com/environment/2019/oct/09/climate-emergency-what-oil-gas-giants-say.

[5] Nick Cunningham, "Oil Industry 'Net-Zero' Pledges Are an Attempt to Delay Climate Action, New Paper Warns," *DeSmog*, April 15, 2021, https://www.desmog.com/2021/04/15/oil-industry-net-zero-pledges-delay-climate-action-study.

[6] Shannon Hall, "Exxon Knew about Climate Change almost 40 years ago," *Scientific American*, October 26, 2015, https://www.scientificamerican.com/article/exxon-knew-about-climate-change-almost-40-years-ago/.

[7] Dana Drugmand, "Polling Shows Growing Climate Concern Among Americans. But Outsized Influence of Deniers Remains a Roadblock," *DeSmog*, Oct 22, 2020, https://www.desmogblog.com/2020/10/22/polling-concern-americans-climate-deniers-exxon.

[8] "Sen. James M Inhofe—Oklahoma, Top Industries 2015-2020," OpenSecrets.org, https://www.opensecrets.org/members-of-congress/james-m-inhofe/industries?cid=N00005582&cycle=2020&type=C.

[9] Heartland Institute, "13th International Conference on Climate Change," https://www.heartland.org/events/events/13th-international-conference-on-climate-change.

[10] See Peter H. Gleick, "Book Review: Bad Science and Bad Arguments Abound in 'Apocalypse Never' by Michael Shellenberger," *Yale Climate Connections*, July 15, 2020, https://yaleclimateconnections.org/2020/07/review-bad-science-and-bad-arguments-abound-in-apocalypse-never/.

[11] Lydia Maria Child, *The Patriarchal Institution: As Described by Members of Its Own Family* (New York: American Anti-Slavery Society, 1860), 5.

[12] John Guttman, "Was John C. Calhoun Proud of Slavery?" HistoryNet.com, https://www.historynet.com/was-john-c-calhoun-proud-of-slavery.htm.

[13] Josiah C. Nott, "Two Lectures on the Natural History of the Caucasian and Negro Races," *The Ideology of Slavery*, 206.

[14] American Petroleum Institute, "The Energy to Power America or Climate Solutions? We Don't Have to Choose," EnergyforProgress.org, https://energyforprogress.org/the-basics/.

[15] Laura Santhanam, "Most Americans Would Pay More to Avoid Using Plastic, Poll Says," *PBS NewsHour*, November 26, 2019,

https://www.pbs.org/newshour/nation/most-americans-would-pay-more-to-avoid-using-plastic-poll-says.

[16] James Henry Hammond, "Letter to An English Abolitionist," *The Ideology of Slavery*, 177.

[17] American Petroleum Institute, "Energy for All," EnergyCitizens.org, https://energycitizens.org/discover/energy-for-all/.

[18] Michael Shellenberger, *Apocalypse Never: Why Environmental Alarmism Hurts Us All* (New York: Harper), 248.

[19] Gayathri Vaidyanathan, "Coal Trumps Solar in India," *Scientific American*, October 19, 2015, https://www.scientificamerican.com/article/coal-trumps-solar-in-india/.

[20] Will Mathis, "Building New Renewables Is Cheaper than Burning Fossil Fuels," *Bloomberg Green*, June 23, 2021, https://www.bloomberg.com/news/articles/2021-06-23/building-new-renewables-cheaper-than-running-fossil-fuel-plants.

[21] For example, see Canadian Commission for Environmental Cooperation, "Renewable Energy as a Hedge Against Fuel Price Fluctuation," 2008, http://www3.cec.org/islandora/fr/item/2360-renewable-energy-hedge-against-fuel-price-fluctuation-en.pdf.

[22] Oxfam, "World's Richest 10% Produce Half of Carbon Emissions While Poorest 3.5 Billion Account for Just a Tenth," December 2, 2015, https://www.oxfam.org/en/press-releases/worlds-richest-10-produce-half-carbon-emissions-while-poorest-35-billion-account.

[23] Jason Hickel, "We Can't Have Billionaires and Stop Climate Change," *The Correspondent*, October 9, 2020, https://thecorrespondent.com/728/we-cant-have-billionaires-and-stop-climate-change/842640975176-f7bab0dc.

[24] James Henry Hammond, *Remarks of Mr. Hammond, of South Carolina, on the Question of Receiving Petitions for the Abolition of Slavery in the District of Columbia* (Washington, DC: Duff Green, 1836), 11.

[25] Isabel Wilkerson, *Caste: The Origins of Our Discontents* (New York: Random House, 2020), 45.

[26] Sammy Roth, "The Fossil Fuel Industry Wants You to Believe It's Good For People Of Color," *Los Angeles Times*, November 23, 2020, https://www.latimes.com/business/story/2020-11-23/clean-energy-fossil-fuels-racial-justice.

[27] Emily Atkin, "The Quiet Campaign to Make Clean Energy Racist," *Heated*, July 2, 2020, https://heated.world/p/the-quiet-campaign-to-make-clean.

[28] National Association for the Advancement of Colored People, "Fossil Fueled Foolery: An Illustrated Primer on the Top 10 Manipulation Tactics of the Fossil Fuel Industry," NAACP.org, April 1, 2019, https://www.naacp.org/latest/new-naacp-report-fossil-fueled-foolery-group-highlights-top-ten-ways-fossil-fuel-companies-fool-public/.

[29] American Petroleum Institute, "The Energy to Power America or Climate Solutions?" https://energyforprogress.org/the-basics/.

[30] Jeffrey Rissman and Robbie Orvis, "Carbon Capture And Storage: An Expensive Option For Reducing U.S. CO2 Emissions," *Forbes*, May 3, 2017, https://www.forbes.com/sites/energyinnovation/2017/05/03/carbon-capture-and-storage-an-expensive-option-for-reducing-u-s-co2-emissions/.

[31] James Henry Hammond, "Plantation Manual," *James Henry Hammond Papers, 1857-1858*, https://memory.loc.gov/ammem/awhhtml/awmss5/d08.html.

[32] See Andrew Delbanco, *The War Before the War: Fugitive Slaves and the Struggle for America's Soul from the Revolution to the Civil War* (New York: Penguin, 2018).

[33] Paul Griffin, *Carbon Majors Report 2017*, Climate Accountability Institute, https://climateaccountability.org/pdf/CarbonMajorsRpt2017%20Jul17.pdf.

[34] Kate Yoder, "Footprint Fantasy," *Grist*, August 26, 2020, https://grist.org/energy/footprint-fantasy/.

[35] Kate Yoder, "Greentrolling: A 'Maniacal Plan' to Bring Down Big Oil," *Grist*, November 19, 2020, https://grist.org/energy/greentrolling-a-maniacal-plan-to-bring-down-big-oil/.

[36] Nathanael Johnson, "Have You Been Doing Environmentalism Wrong?" *Grist*, April 28, 2021, https://grist.org/climate-energy/have-you-been-doing-environmentalism-wrong/.

[37] Andrew Nikiforuk, *The Energy of Slaves: Oil and the New Servitude* (Vancouver, BC: Greystone Books, 2014).

[38] Michael Mann, "Lifestyle Changes Aren't Enough to Save the Planet. Here's What Could," *Time*, September 12, 2019, https://time.com/5669071/lifestyle-changes-climate-change/.

[39] Mary Heglar, "I Work in The Environmental Movement. I Don't Care if You Recycle," *Vox*, June 4, 2019, https://www.vox.com/the-highlight/2019/5/28/18629833/climate-change-2019-green-new-deal.

[40] James Brewer Stewart, "Is the Green New Deal Impossible?"

[41] Evan Andrews, "How Many U.S. Presidents Owned Enslaved People?" History.com, September 3, 2019, https://www.history.com/news/how-many-u-s-presidents-owned-slaves.

[42] University of Houston Digital History, "Answers to the Quiz on Slavery," https://www.digitalhistory.uh.edu/disp_textbook.cfm?smtID=13&psid=3757.

[43] Abraham Lincoln, Speech in Peoria, IL, October 16, 1854, *Lincoln's Writings: The Multimedia Edition*, https://housedivided.dickinson.edu/sites/lincoln/peoria-speech-october-16-1854/.

[44] Senate Republican Policy Committee, "Green New Deal: A Crazy, Expensive Mess," December 11, 2018, https://www.rpc.senate.gov/policy-papers/green-new-deal-a-crazy-expensive-mess.

[45] Marina Andrijevic, *etal.*, "COVID-19 Recovery Funds Dwarf Clean Energy Investment Needs," *Science*, October 16, 2020, https://science.sciencemag.org/content/370/6514/298.

[46] Adele Peters, "The Enormous COVID-19 Recovery Plans Show There's Money to Solve Climate Change," *Fast Company*, October 23, 2020, https://www.fastcompany.com/90567028/the-enormous-covid-19-recovery-plans-show-theres-money-to-solve-climate-change.

[47] Joseph Stiglitz, "The Economic Case for a Green New Deal," *Winning the Green New Deal*, 103.

[48] Manisha Sinha, "The 2020 Election Surpasses All Before it, Except One," *CNN*, October 28, 2020, https://www.cnn.com/2020/10/28/opinions/2020-election-biggest-since-1860-sinha/.

49 Lehrman Institute, "Compensated Emancipation," *Mr. Lincoln and Freedom*, http://www.mrlincolnandfreedom.org/civil-war/congressional-action-inaction/compensated-emancipation/.

Chapter 4

1 Fiona Harvey, "Humanity is Waging War On Nature, Says UN Secretary General," *The Guardian*, December 2, 2020, https://www.theguardian.com/environment/2020/dec/02/humanity-is-waging-war-on-nature-says-un-secretary-general-antonio-guterres.
2 Gwynne Dyer, *Climate Wars: The Fight for Survival as the World Overheats* (New York: Oneworld Publications, 2010), xii.
3 "Abraham Lincoln Inaugurated," History.com, March 2, 2021, https://www.history.com/this-day-in-history/lincoln-inaugurated.
4 Northern states had already started to approve the Crittenden Compromise, part of which is also called the Corwin Amendment. Ironically, only the outbreak of civil war started by slaveowners themselves prevented the likely passage of a Thirteenth Amendment to the U.S Constitution that would have forever protected slavery. Fortunately, the next chance that America had to pass a constitutional amendment on slavery came at the end of the Civil War when attitudes had reversed, away from compromise with slaveowners and towards removing slavery as a cause of sectional conflict in the future. The Thirteenth Amendment that we know today did the very opposite of the Crittenden Compromise: it forever outlawed slavery in the United States. The amendment was approved by Congress in January 1865 and passed by enough states to make it law in December 1865. For more see "A Proposed Thirteenth Amendment to Prevent Secession, 1861," *History Resources*, Gilder Lehrman Institute of American History, https://www.gilderlehrman.org/history-resources/spotlight-primary-source/proposed-thirteenth-amendment-prevent-secession-1861.
5 Michael Mann, *The New Climate War*, 259.
6 Elizabeth Nix, "The Worst Picnic in History Was Interrupted by a War," History.com, August 30, 2018, https://www.history.com/news/worst-picnic-in-history-was-interrupted-by-war.
7 Michael Mann, *The New Climate War*, 258.
8 United Nations, "UN Secretary-General: 'Making Peace with Nature is the Defining Task of the 21st century,'" December 2, 2020, https://unfccc.int/news/un-secretary-general-making-peace-with-nature-is-the-defining-task-of-the-21st-century.
9 Some historians estimate that as many as 850,000 people died on both sides in the Civil War, but the American Battlefield Trust disagrees with that claim, instead favoring an estimate made by a study performed in the late nineteenth century of military records including enlistment rolls, muster rolls, and casualty lists from Union and Confederate records. See "Civil War Casualties," American Battlefield Trust, https://www.battlefields.org/learn/articles/civil-war-casualties.
10 Tom Jones, "Confederates' Lost Cause Still Cripples the South's Economy," *MLK50*, July 27, 2017, https://mlk50.com/2017/07/27/confederates-lost-cause-still-cripples-the-souths-economy/.

[11] Abraham Lincoln, "Second Inaugural Address," *Lincoln: Speeches and Writings 1859-1865* (New York: Library of America, 1989), 686-687.

[12] H.W. Brands, "Ulysses S. Grant: The Man Who Saved the Union," Rancho Mirage Writers Festival, Feb. 17, 2017, https://youtu.be/ldLMw-SFE6E.

[13] David Blight, *Race and Reunion: The Civil War in American Memory* (Cambridge, MA: Harvard University Press, 2001), 306.

[14] David French, *Divided We Fall: America's Secession Threat and How to Restore our Nation* (New York: St. Martin's, 2020), 1-2.

[15] Evan Osnos, "Pulling Our Politics Back from the Brink," *The New Yorker*, Nov. 9, 2020, https://www.newyorker.com/magazine/2020/11/16/pulling-our-politics-back-from-the-brink.

[16] Richard Kreitner, *Break It Up: Secession, Division, and the Secret History of America's Imperfect Union* (New York: Little, Brown and Company, 2020), 238.

[17] Joanne B. Freeman, *The Field of Blood: Violence in Congress and the Road to Civil War* (New York: Farrar, Strauss and Giroux, 2018), xiii.

[18] Manisha Sinha, *The Slave's Cause: A History of Abolition* (New Haven, CT: Yale University Press, 2018), 419.

[19] Henry Mayer, *All on Fire: William Lloyd Garrison and the Abolition of Slavery* (New York: St. Martin's Press, 1998), 121.

[20] For Freedoms email newsletter sent December 16, 2020.

[21] Sunrise Movement, "We Are Wide Awake," https://www.sunrisemovement.org/campaign/wide-awake/.

[22] Frederick Douglass, "Address of Hon. Fred. Douglass, delivered before the National Convention of Colored Men, at Louisville, Ky., September 24, 1883," https://omeka.coloredconventions.org/items/show/554.

[23] CNA Corporation, *National Security and the Threat of Climate Change*, April 2007, 31, https://www.cna.org/cna_files/pdf/national%20security%20and%20the%20threat%20of%20climate%20change.pdf.

[24] Anatol Lieven, *Climate Change and the Nation State: The Case for Nationalism in a Warming World* (New York: Oxford University Press, 2020), 36.

[25] Paul D. Eaton, "National Security and Combating the Climate Crisis Go Hand in Hand," *World War Zero Magazine*, September 9, 2020, https://worldwarzero.com/magazine/2020/09/national-security-the-climate-crisis-go-hand-in-hand/.

[26] Michael Klare, *All Hell Breaking Loose: The Pentagon's Perspective on Climate Change* (New York: Metropolitan Books, 2019), 33.

[27] Henry Fountain and John Schwartz, "Report Provides a Preview of the 'New Arctic'," *New York Times*, December 9, 2020, https://www.nytimes.com/2020/12/09/climate/climate-book-suggestions.html.

[28] Gwynne Dyer, *Climate Wars*, 17.

[29] Michael Klare, *All Hell Breaking Loose*, 36.

[30] "2-Metre Sea Level Rise 'Plausible' by 2100: Study," Phys.org, May 21, 2019, https://phys.org/news/2019-05-metre-sea-plausible.html.

[31] Quoted in Anatol Lieven, *Climate Change and the Nation State*, 19.

[32] Michael Klare, *All Hell Breaking Loose*, 36.

[33] Michael Klare, *All Hell Breaking Loose*, 38.

[34] Neta C. Crawford, "Cutting Military Emissions Is a Matter of National Security," *Undark,* June 18, 2019, https://undark.org/2019/06/18/defense-climate-change.

[35] Henry Mayer, *All on Fire,* 502.

[36] Frederick Douglass, "John Brown, an Address by Frederick Douglass, at The Fourteenth Anniversary of Storer College, Harper's Ferry, West Virginia, May 30, 1881," (Dover, NH: Morning Star Job Printing House, 1881).

[37] Britt Rusert, "The Radical Lives of Abolitionists," *Boston Review,* January 20, 2020, http://bostonreview.net/race/britt-rusert-radical-lives-abolitionists.

[38] Anatol Lieven, *Climate Change and the Nation State,* 7.

[39] Erica Chenoweth and Maria J. Stephan, *Why Civil Resistance Works: The Strategic Logic of Nonviolent Conflict* (New York: Columbia University Press, 2012).

[40] Michael Mann, *The New Climate War,* 232-233.

[41] Roudabeh Kishi and Sam Jones, "Demonstrations & Political Violence In America: New Data for Summer 2020," Armed Conflict Location & Event Data Project, September 2020, https://acleddata.com/2020/09/03/demonstrations-political-violence-in-america-new-data-for-summer-2020/.

Chapter 5

[1] Holly Jackson, *American Radicals,* 195.

[2] Kevin Crowley and Akshat Rathi, "Exxon Mobil's Investment Plan Adds Millions of Tons of Carbon Output, Documents Reveal," *World Oil,* October 5, 2020, https://www.worldoil.com/news/2020/10/5/exxon-mobil-s-investment-plan-adds-millions-of-tons-of-carbon-output-documents-reveal.

[3] Michael Mann, *The New Climate War,* 4-5.

[4] Yohuru Williams, "Why Thomas Jefferson's Anti-Slavery Passage Was Removed from the Declaration of Independence," History.com, June 29, 2020, https://www.history.com/news/declaration-of-independence-deleted-antislavery-clause-jefferson.

[5] Though they could have been hanged as pirates, slave ship captains found ways to evade underfunded British and American naval patrols off the west African coast and deliver kidnapped Africans to buyers in Brazil and Cuba. A handful of illegal slave ships also called at ports in U.S. Southern states right through the Civil War. But there was plenty of guilt to go around. For decades after the slave trade was banned, Northern ports including New York and Boston became clandestine sources of ships for a shadowy slave-trading syndicate called "the Portuguese Company." Yet, the Northern public put most of the blame for the illegal trade on Southern buyers. The Clotilda, which smuggled 110 Africans to the Gulf Coast in 1860, was the last known slave ship to arrive in the United States. The discovery of its wreckage in the Mobile River in Alabama was featured in a *60 Minutes* special in November 2020. Lincoln and the newly formed Republican Party were opposed to the trade wherever it took place, but they made an election issue out of lambasting Southern attempts to defy federal law. See Manuel Barcia and John Harris, "The Slave Trade Continued Long After It Was Illegal—With Lessons for Today," *Washington Post,* December 6, 2020, https://www.washingtonpost.com/outlook/2020/12/06/slave-trade-continued-long-after-it-was-illegal-with-lessons-today/.

[6] Matthew Karp, *This Vast Southern Empire: Slaveholders at the Helm of American Foreign Policy* (Cambridge, MA: Harvard University Press, 2018), 84.

[7] Matthew Karp, *This Vast Southern Empire*, 181-182.

[8] Darren Dochuk, *Anointed with Oil: How Christianity and Crude Made Modern America* (New York: Basic Books, 2019), 30.

[9] Darren Dochuk, *Anointed with Oil*, 1-6.

[10] Daniel Yergin, *The Prize: The Epic Quest for Oil, Money, and Power* (New York: Simon & Schuster, 1992), 244-245.

[11] Darren Dochuk, *Anointed with Oil*, 34-35.

[12] Matthew Karp, *This Vast Southern Empire*, 4-5.

[13] Matthew Karp, *This Vast Southern Empire*, 17.

[14] See Lawrence Wright, *God Save Texas: A Journey into the Soul of the Lone Star State* (New York: Vintage, 2019).

[15] Matthew Karp, *This Vast Southern Empire*, 124.

[16] Matthew Karp, *This Vast Southern Empire*, 220.

[17] Jonathan Horn, *The Man Who Would Not Be Washington: Robert E. Lee's Civil War and His Decision that Changed American History* (New York: Scribner, 2015), 97.

[18] Andrew F. Lang, *A Contest of Civilizations: Exposing the Crisis of American Exceptionalism in the Civil War Era* (Chapel Hill, NC: University of North Carolina Press, 2021), 148.

[19] Tegan Hanlon, "Trump Rushes to Lock in Oil Drilling in Arctic Wildlife Refuge Before Biden's Term," *NPR*, December 3, 2020, https://www.npr.org/2020/12/03/942052004/trump-administration-sets-last-minute-oil-lease-sale-for-arctic-wildlife-refuge.

[20] Rachel Koning Beals, "Bank of America Joins Big U.S. Banks That Won't Finance Oil In The Arctic Refuge Trump Opened to Drilling," *Marketwatch*, December 5, 2020, https://www.marketwatch.com/story/bank-of-america-joins-big-u-s-banks-that-wont-finance-oil-in-the-arctic-refuge-trump-opened-to-drilling-11606843342.

[21] Coral Davenport, Henry Fountain, and Lisa Friedman, "Biden Suspends Drilling Leases in Arctic National Wildlife Refuge," *New York Times*, June 1, 2021, https://www.nytimes.com/2021/06/01/climate/biden-drilling-arctic-national-wildlife-refuge.html.

[22] "Energy/Natural Resources Top Contributors 2019-2020," OpenSecrets.org, https://www.opensecrets.org/industries/indus.php?ind=E.

[23] Pam Martens and Russ Martens, "Charles Koch Attempts an Apology Tour after He and His Father Financed a Political Hate Machine for Six Decades," *Wall Street on Parade*, November 16, 2020, https://wallstreetonparade.com/2020/11/charles-koch-attempts-an-apology-tour-after-he-and-his-father-financed-a-political-hate-machine-for-six-decades/.

[24] President Donald J. Trump, "Promoting Energy Independence and Economic Growth," Executive Order 13783, March 28, 2017, https://www.federalregister.gov/documents/2017/03/31/2017-06576/promoting-energy-independence-and-economic-growth.

[25] David Roberts, "Donald Trump is Handing the Federal Government Over to Fossil Fuel Interests," *Vox*, June 14, 2017, https://www.vox.com/energy-and-environment/2017/6/13/15681498/trump-government-fossil-fuels.

[26] David Roberts, "Donald Trump is Handing the Federal Government Over to Fossil Fuel Interests."

[27] Sharon Kelly, "Climate Deniers Backed Violence and Spread Pro-Insurrection Messages Before, During, and After January 6," *DeSmog*, February 16, 2021, https://www.desmogblog.com/2021/02/16/climate-deniers-messages-support-capitol-insurrection.

[28] Pam Martens and Russ Martens, "Charles Koch Attempts an Apology Tour after He and His Father Financed a Political Hate Machine for Six Decades."

[29] Marcella Mulholland, "How Kansas' Koch Industries Helped Fuel the U.S. Capitol Insurrection," *Kansas Reflector*, January 21, 2021, https://kansasreflector.com/2021/01/21/how-kansas-koch-industries-helped-fuel-the-u-s-capitol-insurrection/.

[30] Ellen M. Gilmer and Stephen Lee, "Biden's Climate Support Could Spawn More Cases Against Big Oil," *Bloomberg Law*, July 22, 2020, https://news.bloomberglaw.com/environment-and-energy/bidens-climate-support-could-spawn-more-cases-against-big-oil.

[31] "Chevron Corp PAC Contributions to Federal Candidates," OpenSecrets.org, https://www.opensecrets.org/political-action-committees-pacs/chevron-corp/C00035006/candidate-recipients/2020.

[32] Abbey Dufoe, "Unpacking Chevron's Call for a Peaceful Transition of Power," *ExxonKnews*, January 8, 2021, https://exxonknews.substack.com/p/unpacking-chevrons-call-for-a-peaceful.

[33] Emily Sanders, "Climate Liability Has Its Day at The Supreme Court (Sort of)," *ExxonKnews*, January 22, 2021, https://exxonknews.substack.com/p/climate-liability-has-its-day-at.

[34] Alexander C. Kaufman, "Solar Expected to Dethrone Coal, Become 'New King Of Electricity,' Global Forecast Finds," *Huffpost*, October, 13, 2020, https://www.huffpost.com/entry/solar-power-coal-energy_n_5f8640fdc5b6e9e76fb847d2.

[35] Hanna Ziady, "BP Will Slash Oil Production by 40% and Pour Billions into Green Energy," *CNN*, August 4, 2020, https://www.cnn.com/2020/08/04/business/bp-oil-clean-energy/index.html.

[36] Javier E. David, "'Beyond Petroleum' No More? BP Goes Back to Basics," *CNBC*, April 23, 2013, https://www.cnbc.com/id/100647034.

[37] Emily Pontecorvo, "Exxon's 'Emission Reduction Plan' Doesn't Call for Reducing Exxon's Emissions," *Grist*, December 15, 2020, https://grist.org/energy/exxons-emission-reduction-plan-doesnt-call-for-reducing-exxons-emissions.

[38] Jonathan Watts, Jillian Ambrose, and Adam Vaughan, "Oil Firms to Pour Extra 7m Barrels Per Day into Markets, Data Shows," *The Guardian*, October 10, 2019, https://www.theguardian.com/environment/2019/oct/10/oil-firms-barrels-markets.

[39] Jonathan Watts, Jillian Ambrose, and Adam Vaughan, "Oil Firms to Pour Extra 7m Barrels Per Day into Markets, Data Shows."

[40] Nick Cunningham, "Oil Industry 'Net-Zero' Pledges Are an Attempt to Delay Climate Action, New Paper Warns," *DeSmog*, April 15, 2020, https://www.desmog.com/2021/04/15/oil-industry-net-zero-pledges-delay-climate-action-study/.

41 Tim Fernholz, "ExxonMobil Made Illegal Deals with Russia When Secretary of State Rex Tillerson Ran the Company," *Quartz,* July 20, 2017, https://qz.com/1034589/exxonmobil-made-illegal-deals-with-russia-when-secretary-of-state-rex-tillerson-ran-the-company/.

42 Jonathan Watts, Jillian Ambrose, and Adam Vaughan, "Oil Firms to Pour Extra 7m Barrels Per Day into Markets, Data Shows."

43 "Why ExxonMobil is Sticking with Oil as Rivals Look to a Greener Future," *Financial Times,* October 27, 2020, https://www.ft.com/content/30ffa51b-2079-400e-84f1-2e45991194c8.

44 Alexander C. Kaufman, "Solar Expected To Dethrone Coal, Become 'New King Of Electricity,' Global Forecast Finds," *Huffpost,* October 12, 2020, https://www.huffpost.com/entry/solar-power-coal-energy_n_5f8640fdc5b6e9e76fb847d2.

45 David Roberts, "Big Oil's Hopes Are Pinned on Plastics. It Won't End Well," *Vox,* October 28, 2020, https://www.vox.com/energy-and-environment/21419505/oil-gas-price-plastics-peak-climate-change.

46 Scott K. Johnson, "U.S. Grid-Battery Costs Dropped 70% Over 3 Years," *ArsTechnica,* October 27, 2020, https://arstechnica.com/science/2020/10/us-grid-battery-costs-dropped-70-over-3-years/.

47 Justin Mikulka, "Major Fossil Fuel PR Group is Behind Europe Pro-Hydrogen Push," *DeSmog,* December 9, 2020, https://www.desmogblog.com/2020/12/09/fti-consulting-fossil-fuel-pr-group-behind-europe-hydrogen-lobby.

Chapter 6

1 Elizabeth Heyrick, *Immediate, Not Gradual Abolition, or, An Inquiry Into The Shortest, Safest, and Most Effectual Means of Getting Rid of West Indian Slavery* (Boston: Isaac Knapp, 1838), 22.

2 Justin Mikulka, "Fossil Fuel Tax Programs to Cut Emissions Lead to Lots of Industry Profit, Little Climate Action," *DeSmog,* April 4, 2021, https://www.desmogblog.com/2021/04/04/fossil-fuel-tax-programs-emissions-climate/.

3 Kris Manjapra, "When Will Britain Face Up to Its Crimes Against Humanity?" *The Guardian,* March 29, 2018, https://www.theguardian.com/news/2018/mar/29/slavery-abolition-compensation-when-will-britain-face-up-to-its-crimes-against-humanity.

4 "Britain and the Slave Trade," U.K. National Archives, http://www.nationalarchives.gov.uk/slavery/pdf/britain-and-the-trade.pdf.

5 Adam Hochschild, *Bury the Chains: Prophets and Rebels in the Fight to Free an Empire's Slaves* (New York: Houghton Mifflin, 2005), 55.

6 "Crude Oil Export Ban," *Ballotpedia,* https://ballotpedia.org/Crude_oil_export_ban.

7 Emily Sanders, "The Big Apple Won't Back Down, and a Maryland County Files Suit," *ExxonKnews,* April 30, 2021, https://exxonknews.substack.com/p/the-big-apple-wont-back-down-and.

[8] Mike Kaye, *1807-2007: Over 200 Years of Campaigning Against Slavery (London: Anti-Slavery International*, 2005), 7-9, http://www.antislavery.org/wp-content/uploads/2017/01/18072007.pdf.

[9] Adam Hochschild, *Bury the Chains*, 160.

[10] Mike Kaye, *1807-2007: Over 200 Years of Campaigning against Slavery*, 11.

[11] Mike Kaye, *1807-2007: Over 200 Years of Campaigning against Slavery*, 11.

[12] Thomas Clarkson, "Diagram of the 'Brookes' Slave Ship," *The History of the Rise, Progress, and Accomplishment of the Abolition of the African Slave-Trade by the British Parliament* (London: R. Taylor and Company, 1808), https://www.bl.uk/collection-items/diagram-of-the-brookes-slave-ship.

[13] Hochschild, Bury the Chains, 155-156.

[14] Davis Guggenheim, *An Inconvenient Truth*, 2006. For a clip of the scene with Gore on the lift, see https://www.youtube.com/watch?v=9tkDK2mZlOo.

[15] National Aeronautics and Space Administration, "Apollo Astronaut Shares Story of NASA's Earthrise Photo," NASA.gov, March 29, 2012, https://www.nasa.gov/centers/johnson/home/earthrise.html.

[16] Daksh Sharma, "Breath," International Photography Awards, November 2016, https://www.photoawards.com/winner/zoom.php?eid=8-133709-16.

[17] Adam Hochschild, *Bury the Chains*, 151.

[18] Adam Hochschild, *Bury the Chains*, 159-160.

[19] Adam Hochschild, *Bury the Chains*, 185.

[20] Nick Cunningham, "Oil Industry Inflates Job Impact From Biden's New Pause on Drilling on Federal Lands," *DeSmog*, January 27, 2021, https://www.desmogblog.com/2021/01/27/oil-inflates-job-impact-biden-pause-drilling-federal-lands.

[21] Adam Hochschild, *Bury the Chains*, 154.

[22] Robert Stein, "The Revolution of 1789 and the Abolition of Slavery," *Canadian Journal of History*, Winter 2016, https://www.utpjournals.press/doi/abs/10.3138/cjh.17.3.447.

[23] Mike Kaye, *1807-2007: Over 200 Years of Campaigning against Slavery*, 15.

[24] Michael Brune, "The Sierra Club and Natural Gas," SierraClub.org, February 2, 2012, https://www.sierraclub.org/michael-brune/2012/02/sierra-club-and-natural-gas.

[25] Adam Hochschild, *Bury the Chains*, 343.

[26] Kris Manjapra, "When Will Britain Face Up to Its Crimes Against Humanity?" Adam Hochschild gives a much smaller figure, about $2.2 billion, in *Bury the Chains*, 347.

[27] Richard Conniff, "What's Wrong with Putting a Price on Nature?" *Yale Environment 360*, October 18, 2012, https://e360.yale.edu/features/ecosystem_services_whats_wrong_with_putting_a_price_on_nature.

[28] Matthew Brown, "Fact Check: United Kingdom Finished Paying off Debts to Slave-Owning Families in 2015," *USA Today*, June 30, 2020, https://www.usatoday.com/story/news/factcheck/2020/06/30/fact-check-u-k-paid-off-debts-slave-owning-families-2015/3283908001/.

[29] Sanchez Manning, "Britain's Colonial Shame: Slave-Owners Given Huge Payouts After Abolition," *The Independent*, February 24, 2013,

https://www.independent.co.uk/news/uk/home-news/britain-s-colonial-shame-slave-owners-given-huge-payouts-after-abolition-8508358.html.

30 Richard Barker, "What Abolition of Slavery Tells Us about Climate Change," TEDx, London Business School, June 18, 2018, https://youtu.be/3NI-j8sDKno.

31 Richard Barker, "What Abolition of Slavery Tells Us about Climate Change."

32 Richard Barker, "What Abolition of Slavery Tells Us about Climate Change."

33 Richard Barker, "What Abolition of Slavery Tells Us about Climate Change."

34 Richard Barker, interview with the author, September 26, 2019.

35 Richard Barker, interview with the author, September 26, 2019.

36 Richard Barker, interview with the author, September 26, 2019.

37 Shannon Osaka, "Biden is Canceling Fossil Fuel Subsidies. But He Can't End Them All," *Grist,* January 28, 2021, https://grist.org/politics/biden-is-eliminating-fossil-fuel-subsidies-but-he-cant-end-them-all/.

38 Justin Mikulka, "Fossil Fuel Tax Programs to Cut Emissions Lead to Lots of Industry Profit, Little Climate Action," *DeSmog,* April 4, 2021, https://www.desmogblog.com/2021/04/04/fossil-fuel-tax-programs-emissions-climate/.

39 Jeff Brady, "Exxon Lobbyist Caught on Video Talking About Undermining Biden's Climate Push," *NPR,* July 1, 2021.

Chapter 7

1 See "Milestones in Climate Science," *Skeptical Science,* https://skepticalscience.com/graphics.php?g=59.

2 House of Representatives Energy and Commerce Committee Republicans, "'Coal Keeps the Lights On'," July 25, 2013, https://republicans-energycommerce.house.gov/news/blog/coal-keeps-lights/.

3 Mary Virginia Orna, *etal.,* "Chemistry Feeds the World," *Chemistry's Role in Food Production and Sustainability: Past and Present ACS Symposium Series* (Washington, DC: American Chemical Society, 2019), https://pubs.acs.org/doi/pdf/10.1021/bk-2019-1314.ch001.

4 PlasticsMakeItPossible.com, https://www.plasticsmakeitpossible.com/.

5 Alex Epstein, *The Moral Case for Fossil Fuels* (New York: Portfolio, 2014).

6 Union of Concerned Scientists, "Scientists React to Fossil Fuel Ads," April 11, 2008, https://www.ucsusa.org/resources/scientists-react-fossil-fuel-ads.

7 Stop the Money Pipeline, "About the Movement," https://stopthemoneypipeline.com/about-2/.

8 This phrase sounds less poetic when you know that Truth added it to mark the cards for legal reasons as an original piece of intellectual property created by her at a time when photographers regularly retained the rights to reproduce and sell copies of photos of their subjects.

9 Quoted in the *Archive for Research into Archetypal Symbolism,* "What Are Archetypes?" ARAS.org, https://aras.org/about/what-are-archetypes.

10 Kendra Cherry, "The 4 Major Jungian Archetypes," *Verywell Mind,* June 30, 2020, https://www.verywellmind.com/what-are-jungs-4-major-archetypes-2795439.

11 Andrew Delbanco, *The War Before the War,* 2.

12 Champions of the Earth, "Robert Bullard—Lifetime Achievement Award," https://www.unep.org/championsofearth/laureates/2020/robert-bullard.

13 Matthew Pratt Gutterl, "Frederick Douglass's Faith in Photography," *The New Republic*, November 2, 2015, https://newrepublic.com/article/123191/frederick-douglasss-faith-in-photography.

14 Frederick Douglass, "What to the Slave Is the Fourth of July?" July 5, 1852.

15 Michael Mann, *The New Climate War*, 254.

16 Ken Ellingwood, "Elijah Lovejoy Faced Down Violent Mobs to Champion Abolition and the Free Press," *History News Network*, May 2, 2021, https://historynewsnetwork.org/article/180090.

17 Steve Hanley, "Bill McKibben Calls FBI Tracking Of Environmental Activists 'Contemptible'," *CleanTechnica*, December 13, 2018, https://cleantechnica.com/2018/12/13/bill-mckibben-calls-fbi-tracking-of-environmental-activists-contemptible/.

18 Bill McKibben, "Money Is the Oxygen on Which the Fire of Global Warming Burns," *The New Yorker*, September 17, 2019, https://www.newyorker.com/news/daily-comment/money-is-the-oxygen-on-which-the-fire-of-global-warming-burns.

19 Bill McKibben, "Money Is the Oxygen on Which the Fire of Global Warming Burns."

20 Charles Sumner, *The Barbarism of Slavery: Speech of Hon. Charles Sumner, on the Bill for the Admission of Kansas as a Free State, in the United States Senate, June 4, 1860* (New York: The Young Men's Republican Union, 1863), iv.

21 Michael Mann, *The New Climate War*, 83.

22 The moniker for Al Gore as "America's First Political Climate Action Hero" came from Nathanael Bradford, at the Markkula Center for Applied Ethics, https://www.scu.edu/environmental-ethics/environmental-activists-heroes-and-martyrs/al-gore.html.

23 Al Gore, *Earth in the Balance: Ecology and the Human Spirit*, (New York: Rodale, 2006).

24 Al Gore, *The Assault on Reason: Our Information Ecosystem, from the Age of Print to the Age of Trump Paperback* (New York: Bloomsbury, 2017).

25 "The Nobel Peace Prize 2007," NobelPrize.org, https://www.nobelprize.org/prizes/peace/2007/summary/.

26 "Al Gore—Nobel Lecture," NobelPrize.org, December 10, 2007, https://www.nobelprize.org/prizes/peace/2007/gore/26118-al-gore-nobel-lecture-2007/.

27 James Fallows, "The Planet-Saving, Capitalism-Subverting, Surprisingly Lucrative Investment Secrets of Al Gore," *The Atlantic*, November 2015, https://www.theatlantic.com/magazine/archive/2015/11/the-planet-saving-capitalism-subverting-surprisingly-lucrative-investment-secrets-of-al-gore/407857/.

28 Al Gore, "The Case for Optimism on Climate Change," TED2016, February 2016, https://www.ted.com/talks/al_gore_the_case_for_optimism_on_climate_change/.

29 Bob Dylan, *Chronicles: Volume 1* (New York: Simon & Schuster, 2004).

30 Thaddeus Stevens, *The Selected Papers of Thaddeus Stevens*, Vol. 1, edited by Beverly Wilson Palmer and Holly Byers Ochoa (Pittsburgh: University of Pittsburgh Press, 1997), 248.

[31] The phrase "40 acres and a mule" is most closely associated with Union General William T. Sherman's Special Field Order No. 15, issued on Jan. 16, 1865, four days after a meeting Sherman and Secretary of War Edwin Staunton held with 20 Black leaders in Savannah, Georgia. Stevens was not involved in this specific wartime order but Stevens along with Charles Sumner had been pushing in Congress for land distribution to freed slaves during the war and they continued to promote the idea afterwards.

[32] Fawn Brodie, *Thaddeus Stevens: Scourge of the South* (New York: W.W. Norton, 1966), 180.

[33] Thaddeus Stevens, *The Selected Papers of Thaddeus Stevens*, Vol. 1, 154.

[34] Thaddeus Stevens, *The Selected Papers of Thaddeus Stevens*, Vol. 1, 118.

[35] Thaddeus Stevens, *The Selected Papers of Thaddeus Stevens*, Vol. 1, 120.

[36] Vivian Wang, "Alexandria Ocasio-Cortez: A 28-Year-Old Democratic Giant Slayer," *New York Times*, June 27, 2018, https://www.nytimes.com/2018/06/27/nyregion/alexandria-ocasio-cortez.html.

[37] Nicole Gaudiano, "On Her First Day of Orientation on Capitol Hill, Alexandria Ocasio-Cortez Protests in Pelosi's Office," *USA Today*, November 13, 2018, https://www.usatoday.com/story/news/politics/2018/11/13/alexandria-ocasio-cortez-nancy-pelosi/1987514002/.

[38] David Roberts, "The Green New Deal, Explained," *Vox*, March 30, 2019, https://www.vox.com/energy-and-environment/2018/12/21/18144138/green-new-deal-alexandria-ocasio-cortez.

[39] Associated Press, "Study: Fox News is Obsessed with Alexandria Ocasio-Cortez," *USA Today*, April 14, 2019, https://www.usatoday.com/story/life/tv/2019/04/14/study-fox-news-obsessed-alexandria-ocasio-cortez/3466493002/.

[40] Stephanie Eckhardt, "Alexandria Ocasio-Cortez on the Capitol Riot: 'I Thought I Was Going to Die'," *W Magazine*, February 2, 2021, https://www.wmagazine.com/story/aoc-ocasio-cortez-capitol-abuse.

[41] Fergus M. Bordewich, *Congress at War: How Republican Reformers Fought the Civil War, Defied Lincoln, Ended Slavery, and Remade America* (New York: Alfred A. Knopf, 2020), 57.

[42] Bob Inglis, "A Foreseeable Fire: A Steady Diet of Red Meat Turned the Tea Party into Trumpism," *USA Today*, January 10, 2021, https://www.usatoday.com/story/opinion/2021/01/10/tea-party-protests-trump-bob-inglis-column/6594986002/.

[43] Eric Althoff, "Bob Inglis Advocates Action to Fight Climate Change in Break from Republican Party," *Washington Times*, March 24, 2015, https://www.washingtontimes.com/news/2015/mar/24/bob-inglis-advocates-action-to-fight-climate-chang/.

[44] Sarah Hopkins Bradford, *Scenes in the Life of Harriet Tubman* (Auburn, NY: W. J. Moses, Printer, 1869), 23.

[45] Erica Armstrong Dunbar, *She Came to Slay: The Life and Times of Harriet Tubman* (New York: Simon & Schuster, 2019), 60.

[46] Sarah Hopkins Bradford, *Scenes in the Life of Harriet Tubman*, 21.

[47] Harriet Beecher Stowe, *The Annotated Uncle Tom's Cabin*, 2007.

[48] Colin Woodard, *Union: The Struggle to Forge the Story of United States Nationhood* (New York: Viking, 2020), 128.

[49] Amy Dru Stanley quoted by Henry Louis Gates in *The Annotated Uncle Tom's Cabin*, xiv-xv.

[50] W.E.B. DuBois, *Black Reconstruction in America 1860-1880* (New York: The Free Press, 1998), 83.

[51] H. De Marsan, "Glory Hallelujah. No. 3. or New John Brown Song," *American Song Sheets* in the Library of Congress Rare Books and Special Collections.

[52] Amanda Gorman, "Earthrise," *eePro*, The North American Association for Environmental Education, January 9, 2019, https://naaee.org/eepro/blog/earthrise-poem-amanda-gorman.

[53] Michael Mann, *The New Climate War*, 254.

[54] Jesse Hutchinson Jr., "Get Off the Track! (A Song for Emancipation)," 1844.

Conclusion

[1] W.E.B. DuBois, *Black Reconstruction in America 1860-1880*, 122.

[2] It took white historians a few decades to catch up with the conclusions that W.E.B. DuBois came to in 1935 in *Black Reconstruction*, but the work of Eric Foner, especially *Reconstruction: America's Unfinished Revolution, 1863-1877* published in 1988 with an updated edition in 2014 (New York: HarperPerennial), represents a more positive, revisionist view of Reconstruction. More recently, the revisionist view can be found in Ron Chernow's *Grant* (New York: Penguin, 2017).

[3] Eric Foner, *Reconstruction: America's Unfinished Revolution 1863-1877*, 280.

[4] W.E.B. DuBois, *Black Reconstruction in the South*, 30.

[5] Eric Foner, *Reconstruction*, 603.

[6] William J. Barber, "Rev. Barber: We Are Witnessing the Birth Pangs of a Third Reconstruction," *ThinkProgress*, December 15, 2016, https://archive.thinkprogress.org/rev-barber-moral-change-1ad2776df7c/.

[7] Mark Z. Jacobson *etal.*, "Impacts of Green New Deal Energy Plans on Grid Stability, Costs, Jobs, Health, and Climate in 143 Countries," *OneEarth*, December 20, 2019, https://www.cell.com/one-earth/fulltext/S2590-3322(19)30225-8#%20.

[8] Elizabeth Kolbert, "Biden's Jobs Plan Is Also a Climate Plan. Will It Make a Difference?" *The New Yorker*, April 4, 2021, https://www.newyorker.com/magazine/2021/04/12/bidens-jobs-plan-is-also-a-climate-plan-will-it-make-a-difference.

[9] Jariel Arvin, "How to Spot the Tricks Big Oil Uses to Subvert Action on Climate Change," *Vox*, February 1, 2021, https://www.vox.com/22260311/oil-gas-fossil-fuel-companies-climate-change.

[10] "How Can the Media Cover Climate Change Better?" *The Brian Lehrer Show*, September 19, 2019, https://youtu.be/tWzfq0Qy5mk.

[11] Homepage, EndClimateSilence.org, https://www.endclimatesilence.org/.

[12] https://www.endclimatesilence.org/.

[13] Jim Wallis, *God's Politics: Why the Right Gets It Wrong and the Left Doesn't Get It* (San Francisco: HarperSanFrancisco), 347-348.

Made in the USA
Middletown, DE
31 December 2021

57371193R00166